Archimedes

NEW STUDIES IN THE HISTORY AND PHILOSOPHY OF SCIENCE AND TECHNOLOGY

VOLUME 60

Archimedes has three fundamental goals: to further the integration of the histories of science and technology with one another; to investigate the technical, social and practical histories of specific developments in science and technology; and finally, where possible and desirable, to bring the histories of science and technology into closer contact with the philosophy of science.

The series is interested in receiving book proposals that treat the history of any of the sciences, ranging from biology through physics, all aspects of the history of technology, broadly construed, as well as historically-engaged philosophy of science or technology. Taken as a whole, Archimedes will be of interest to historians, philosophers, and scientists, as well as to those in business and industry who seek to understand how science and industry have come to be so strongly linked.

Submission / Instructions for Authors and Editors: The series editors aim to make a first decision within one month of submission. In case of a positive first decision the work will be provisionally contracted: the final decision about publication will depend upon the result of the anonymous peer-review of the complete manuscript. The series editors aim to have the work peer-reviewed within 3 months after submission of the complete manuscript.

The series editors discourage the submission of manuscripts that contain reprints of previously published material and of manuscripts that are below 150 printed pages (75,000 words). For inquiries and submission of proposals prospective authors can contact one of the editors:

Please find on the top right side of this webpage a link to our *Book Proposal Form.*

More information about this series at http://www.springer.com/series/5644

Wolfgang Lefèvre

Minerva Meets Vulcan: Scientific and Technological Literature – 1450–1750

Springer

Wolfgang Lefèvre
Max Planck Institute for the History of Science
Berlin, Berlin, Germany

ISSN 1385-0180 ISSN 2215-0064 (electronic)
Archimedes
ISBN 978-3-030-73084-0 ISBN 978-3-030-73085-7 (eBook)
https://doi.org/10.1007/978-3-030-73085-7

This Springer imprint is published by the registered company Springer Nature Switzerland AG
The registered company address is: Gewerbestrasse 11, 6330 Cham, Switzerland

Acknowledgements

This book could not have been written without the opportunity to work at the Max Planck Institute of the History of Science, Berlin, with its rich resources. I am very grateful for this unique opportunity and would like to thank especially Jürgen Renn, director of Department I, who supported my work in every way – not least through the intellectually inspiring climate of his department. This stimulating and challenging climate is essentially due to his theoretical approach to the history of science in the frame of an historical epistemology. I hope he can see my book as a contribution to this way of doing history of science and as a reward for his generosity.

My book greatly benefited from discussions with several of the department's research colleagues, in particular Antonio Becchi (London), Jochen Büttner, Ursula Klein, Marcus Popplow (Karlsruhe), and Matthias Schemmel who read and commented on drafts of chapters providing substantial critiques and valuable recommendations. Equally valuable were the reports of the two peer reviewers whom I wish to thank on this way.

At the Max Planck Institute, I always found encouraging support not only by and among its scientific members. I have particularly appreciated the mentoring by Lindy Divarci, editorial manager of Department I, and the manifold services and support of the institute's library organized and provided by its former and current directors, Urs Schoepflin and Esther Chen, as well as by its always highly motivated team of librarians – Sabine Bertram, Urte Brauckmann, Ellen Garske, Ralf Hinrichsen, Ruth Kessentini, Beate MacPhail, Anke Pietzke, and Matthias Schwerdt.

Last but not least, I would like to thank Jed Z. Buchwald for his readiness to enlist this book in his Springer series Archimedes – New Studies in the History and Philosophy of Science and Technology.

Contents

Chapter 1
Introduction

1.1 Science and Technology

This book investigates the relations between the developments of technology and science in the West in the early modern era, that is, the period from the late Middle Ages up to the eighteenth century. It investigates these relations – ranging from those between unrelated parallel courses of development to several forms of interplay and exchange, and then to forms of partial blending – as they are manifested both in the scientific and technological literature.

As we know, speaking of "science" and "technology" in relation to the early modern age is somewhat anachronistic and needs clarification at the outset. As regards "science," it has been shown that modern natural science, which is developing as an increasingly differentiated totality of scientific disciplines, did not emerge until the period from the mid-eighteenth to the mid-nineteenth century. Before this formation of modern scientific disciplines, structures and laws of nature had been studied and conceptualized by and in the frame of traditional domains of learning, namely *natural history*, *mixed mathematics* (mechanics, hydrostatics, optics, astronomy, music theory, etc.) and particularly *natural philosophy*.[1] Thus, when speaking in the present book about the relations between science and technology in the early modern period, we specifically mean the relations of the latter to one of these premodern domains of learning.

As regards "technology," things are more complicated. The term, originally an ancient technical term referring to a systematic treatment of grammar and rhetoric, first came to mean knowledge of the arts only at the turn of the seventeenth century, that is, almost at the end of the period studied in this book. And it was not until 1777 that the German professor of economy Johann Beckmann (1739–1811) gave a more precise definition of the kind of knowledge indicated by "technology" (*Technologie*).

[1] See, for instance, Cunningham and Perry (1993).

W. Lefèvre, *Minerva meets Vulcan: Scientific and Technological Literature – 1450-1750*, Archimedes 60, https://doi.org/10.1007/978-3-030-73085-7_1

Beckman used the term in the context of German cameralistic endeavors when advocating technology as systematically descriptive knowledge about materials, techniques, and tools employed in the arts and crafts of his time. Finally, in the course of the nineteenth century, and along with the continuing expansion of training facilities for aspiring practical experts, "technology" assumed the additional meaning of knowledge of the principles or laws utilized by these means of production. Only then did "technology" also refer to a special branch of science, the engineering sciences taught at technical universities.[2]

In accordance with the eighteenth-century meaning of the term "technology," in the following we will regard such early modern literature as "technological literature" that transmits practitioners' or experts' knowledge about techniques, materials, and tools employed in the respective contemporary trades and industries.[3]

Our focus on "technological literature," that is on "codified knowledge", means to exclude or ignore "embodied" knowledge to a certain degree – "embodied" knowledge about materials, techniques, and tools characteristic of the crafts, which accounts for a wealth of early modern technological knowledge.[4] In other words, excluded is an essential part of "practical knowledge", that is, as Jürgen Renn put it,

the knowledge resulting from the experiences of specially trained practitioners. It is generated from the pursuit of a special task or the use of specific tools and is characteristic of all kinds of craftsmanship […] It has been transmitted, for long historical periods, as part of the transmission of professional skills.[5]

In this book we will only address this knowledge if and insofar as it is part and parcel of advanced technological fields such as architecture or mining.[6]

---

[2] Johann Beckmann: *Anleitung zur Technology, oder zur Kentniß der Handwerke, Fabriken und Manufacturen, vornehmlich derer, die mit der Landwirthschaft, Polizey und Cameralwissenschaft in nächster Verbindung stehn.* Göttingen: Vandenhoeck, 1777. Unlike in German and other European languages, a distinction between technique and technology did not exist in English. Rather, even today, the term can refer both to the techniques, processes, and tools used in the production of goods or services and to knowledge about these means of production. Furthermore, as regards the latter meaning, "technology" can simply refer to expert knowledge of some of these means or to the special case of the engineering sciences taught at technical universities. For the still unclear meaning of "technology" in English, see, for instance, Schatzberg (2006).

[3] For the problems as well as the requisite care in using the notion "expert" for historical actors of the early modern period, see, for instance, Ash (2019).

[4] For the notions of "embodied" and "codified" knowledge as well as for the significance of both for understanding technology-related knowledge in the early modern period, see, for instance, Popplow (2015).

[5] Renn (2020) 410.

[6] The significance of this practical knowledge, not only for the arts and crafts but also for advanced trades and industries and furthermore not least for "codified" technological knowledge, can hardly be overestimated. That is why the early modern artisan and his workshop became the subject of a rich literature in the last two or three decades. This literature covers the thematic spectrum from the general relationship between making and knowing, e.g. Marchand (2011), to specific practices of knowledge production, e.g. Smith and Schmidt (2008) or Dupré (2017), focuses on workshops as hotbeds of knowledge generation, e.g. Roberts et al. (2007), or on exemplary artisans, e.g. Smith (2004) or Nummedal (2007). In this book, we will refer to several aspects of this subject that concern the relationship between technological and scientific literature – see below Sect. 1.2 of this introduction.

The "codified" technological knowledge under investigation in this book, that is the early modern technological literature examined here, is not just craftsmen's knowledge written down. Rather, it differs from this "embodied" knowledge in several ways. First, its authors were very rarely craftsmen. They were usually persons familiar with advanced technologies that developed alongside – though not completely detached from – the world of ordinary crafts, such as mining, construction, weaponry, etc. In contrast to the oral instructions by which "embodied" knowledge was transmitted, technological literature – treatises, drawings, diagrams, mathematical tables, and so on – also imparted a more general, not locally confined "formalization" or standardization to the knowledge transmitted. This "codified" technological knowledge is, furthermore, of particular interest in view of the relationship between technological and scientific knowledge in the early modern period since it was open to incorporating rules (not theories) borrowed from fields of learned knowledge, e.g. from Euclidean geometry. In a mature stage it even facilitated attempts at conceptualizing the characteristics of materials, procedures, and tools employed in such advanced fields of production.

Among the several current definitions of "technology," one is of particular interest for our investigation because it addresses and directly fixes the relationship between science and technology by understanding the latter as an application of the former. To give some of the many possible examples: Technology is

- *the study and knowledge of the practical, especially industrial, use of scientific discoveries.* (Cambridge Dictionary)
- *the application of scientific knowledge for practical purposes, especially in industry.* (Lexico Oxford)
- *refers to methods, systems, and devices which are the result of scientific knowledge being used for practical purposes.* (Collins Dictionary)

True, these definitions refer to modern technological sciences, not to early modern technology. The linear model of the relationship between science and technology they presuppose was and is, however, usually taken as universally valid irrespective of its inapplicability to the early modern period. Up to the eighteenth century there were no scientific theories like nineteenth-century electromagnetism that practitioners could have "applied" and transformed into techniques and procedures or tools and other means of production. Moreover, in this historical period we usually encounter the opposite case that theories are conceptualizations of existing technical items. To give a classic example: The steelyard, the balance with unequal arms, was not an "application" of Archimedes' statics; rather, classical statics was a result of Archimedes' conceptualization of mundane devices like the steelyard.[7]

Apart from these historical considerations, this linear model can be and has been challenged as regards the relationship between modern technological and natural

[7] See, for example, Lefèvre (2001) and Damerow et al. (2002).

sciences. Edwin Layton did this momentously half a century ago when he replaced the linear model with an "interactive model" (Channell) of science and technology.[8]

Layton saw the communities of science and technology that developed in the course of the nineteenth century as "mirror-image twins." As to the shared features of these *twins*, Layton emphasized the similarity of methods as well as of values and institutions in the natural and the technological sciences.[9] And as regards the differences between these *twins*, he emphasized particularly the different role played by fundamental theories on each side.[10] Accordingly, he conceived the relationship between the *twins* as a symmetric one on this level of development of science and technology and, indeed, advocated an interactive model of science and technology, arguing that "information can be transferred in either direction."[11]

Layton gave an example for his description of the *mirror-image twins*: the emergence of the engineering sciences in the United States in the last decades of the nineteenth century – a process he assessed as a "scientific revolution." "American technology," he wrote,

> *went through a scientific revolution in the 19th century. Technological knowledge was uprooted from its matrix in centuries-old craft traditions and grafted onto science. The technological community, which in 1800 had been a craft affair but little changed since the middle ages, was reconstructed as mirror-image twin of the scientific community.*[12]

---

[8] Layton (1971). For the discussions of Layton's article, see, for instance, Channell (1989) and (2009).

[9] "The engineering sciences, by 1900, constituted a complex system of knowledge, ranging from highly systematic sciences to collections of 'how to do it' rules in engineering handbooks. Some, like the strength of materials and hydraulics, built directly on science; they were often classed as branches of physics. Others, such as the kinematics of mechanisms, evolved from engineering practice. In either case, their development involved the adoption by engineers of the theoretical and experimental methods of science, along with many of the values and institutions associated with their use. By 1900 the point of origin made little difference; the engineering sciences constituted a unity. Those derived from practice took on the qualities of a science in their systematic organization, their reliance on experiment, and in the development of mathematical theory. At the same time, sciences like the strength of materials gradually diverged from physics, assuming the characteristics of an autonomous technological science." (Layton (1971) 567f.)

[10] "In the case of mirror-image twins there is a subtle but irreconcilable difference which is expressed as a change in parity. Between the communities of science and technology there was a switch of values analogous to a change in parity. One way of putting the matter would be to note that while the two communities shared many of the same values, they reversed their rank order. In the physical sciences the highest prestige went to the most abstract and general – that is to the mathematical theorists from Newton to Einstein. Instrumentation and applications generally ranked lowest. In the technological community the successful designer or builder ranked highest, 'mere' theorist the lowest. These differences are inherent in the ends pursued by the two communities: scientists seek to know, technologists to do. These values influence not only the status of occupational specialists, but the nature of the work done and the 'language' in which that work is expressed." (Ibid. 576) Another noteworthy difference Layton pointed to concerns the implementation of mathematical methods. Whereas physicists normally admitted only rigorous mathematical methods, academic engineers relied more on approximations and less rigorous graphical methods.

[11] Ibid. 578.

[12] Ibid. 562.

Whether or not this statement applies to the history of "American technology," it certainly fails to hold true for the historical development of technology in Europe, where the technological community of 1800 was in no way just a "craft affair but little changed since the Middle Ages." This is well known in the case of France where the formation of modern engineering sciences was underway at the time. The famous *École polytechnique* in Paris was already 6 years old in 1800. Moreover, the foundation in 1794 of this central educational institution for aspiring engineers and other technical experts by Lazard Carnot (1753–1823) and Gaspard Monge (1748–1818) was a step in the reform of an already existing cluster of special engineering schools which had arisen since the turn of the sixteenth century and included several *Écoles militaires* (since 1719), the *École des ponts et chaussées* (founded in 1747), and the *École de mines* (founded in 1783).[13]

Less well known are similar developments in some German-speaking states in the second half of the eighteenth century, particularly in Prussia, Saxony, and the countries of the Hapsburg monarchy. Here, too, educational institutions were founded for training aspiring technical experts in various practical fields – mining academies in Freiberg (Saxony) and Schemnitz (in present-day Slovakia), a building academy (*Bauakademie*) and an industrial academy (*Gewerbeakademie*) in Prussia, etc. The main motive of these initiatives, which must be seen against the background of the cameralistic policy of several governments in continental Europe as well as of the emergent Industrial Revolution, was the increasing need for competent civil servants in the growing technical departments of the state bureaucracy. The view of science pursued in these endeavors was the concept of "useful sciences" (*nützliche Wissenschaften*).[14]

Ursula Klein has shown that the concept as well as the implementation of these useful sciences at the turn of the eighteenth century anticipated major features of Layton's engineering sciences. Not only was useful knowledge the ultimate aim of both but natural-scientific and technological inquiry were also interdependent or even converged in both cases since the objects of their inquiry overlapped. As regards research, the useful sciences organized research, established research laboratories, and picked up research methods from the natural sciences like Layton's engineering sciences a century later. Finally, the practitioners of both useful science and engineering science eschewed the type of high theory elaborated in natural philosophy and in physics, respectively.[15]

These beginnings of modern engineering sciences on the European continent in the course of the eighteenth century were certainly a milestone in the history of the fabric of sciences in the West – a milestone, not a "scientific revolution." In the eighteenth century, neither the French nor the Prussian technological community was "a craft affair but little changed since the middle ages," as Layton put it with

---

[13] See, for instance, Belhoste (2003), Alder (1997), Picon (1992), and Aguillon (1889).

[14] For the Prussian building academy and the industrial academy, see, for instance, Dobbert (1899) and Kahlow (2000); for the mining academies, see Sect. 6.6 of the chapter on mining science.

[15] Klein (2020), esp. chap. 17.

regard to "American technology." Thus, the incipient engineering sciences in France or Prussia did not and were not obliged to "uproot" the existent technology "from its matrix in centuries-old craft traditions and graft it onto science." Rather, these beginnings of modern engineering sciences crowned technological knowledge and conceptions developed in connection with several new and advanced sectors of production alongside the ordinary crafts that had arisen since the fifteenth century. It is these developments and their relationship with learned/scientific knowledge that this book will investigate.

1.2 Practical and Learned Knowledge in the Early Modern Period

Among present-day historians of science, it hardly remains controversial that contact and exchange between educated and practical knowledge played a significant role in the development of the natural sciences and technology in early modern Europe. Several paths for such exchange arose from the late Middle Ages onward, notwithstanding the fact that large parts of the worlds of learning and mundane practices remained out of touch with each other up to the eighteenth century, as is notoriously the case for most of the scholarship cultivated at the universities as well as the majority of crafts. The emergence of more closely related developments in some fields of practical and learned knowledge was essentially facilitated by the formation of an economy of knowledge that fostered contacts and exchange between the two worlds.

As regards the notion of "economy of knowledge," I follow Jürgen Renn, who has defined this economy as the "ensemble of practices and institutions by which societies produce and reproduce knowledge."[16] The formation of this economy in the early modern period induced and was conversely shaped by several instances or entities mediating between educated and practical knowledge – *personae* (clerics, physicians, engineers/artists/architects, "hybrid experts," etc.); *venues* (monasteries, courts, cities, "trading zones," state administrations, etc.); *institutions of knowledge transmission* (apprenticeships, the journeyman system, schools, academies, etc.); *media of knowledge transmission and exchange* (oral instruction, travelling, manuscripts, printed texts, pictures, diagrams, tables, etc.). Several aspects of these instances are of particular interest for our investigations and should therefore be briefly addressed.[17]

[16] See Renn (2020), chap. 8, especially p. 143.

[17] Each of the four instances of the early modern economy of knowledge – *personae*, *venues*, *institutions of transmission*, and *media* – merit portrayal in a chapter of their own or even a book. In this introductory chapter we must confine ourselves – except for the discussion of one controversy – to some key words and facts of particular interest for our investigation topic without going into details of the literature. As far as I know there are no suitable monographs on any of these instances that could be recommended for "further reading."

(1) As regards *personae*, a rich literature on this topic appeared in the last two or three decades. I will not discuss this further here.[18] Instead, I will recall a highly charged ideological controversy of the 1950s, at the height of the Cold War, about the role of artisans and scholars in the Scientific Revolution – the controversy over the "Scholar-and-Craftsman" thesis. In this controversy, A. Rupert Hall had rightly criticized the view "that sees the new scientist of the seventeenth century as a sort of hybrid between the older natural philosopher and the craftsman."[19] This view had, however, never been advocated by Robert Merton, Hall's chosen target in his campaign against vulgar Marxist narratives of the history of science and particularly of the Scientific Revolution which, in Hall's view, ought to be seen as the "formation of the modern scientific attitude" characteristic of the West.[20]

This controversy is recalled here because of Merton's accurate view of the interchange between learned and practical knowledge in the early modern period, which can be taken as a point of departure for our investigations. Merton was not thinking of ordinary craftsmen when speaking of exchanges between learned men and practitioners but of a very specific group of practitioners – a group that will figure prominently in this book: engineers, gunners, land surveyors, instrument makers, and the like: technical experts who distinguished themselves from ordinary craftsmen not only by their special practical knowledge but also by their investigative ambitions and enterprises.[21] And as regards the counterparts of these experts, Merton was not thinking of traditional "scholars," but rather of men of the incipient modern sciences, who approached technical experts and inventors and emphasized the role played by experience in the acquisition of knowledge.

Moreover, Merton also noticed the occasional hybridization of the roles of the practical expert and the scientist, that is the emergence of the figure of a hybrid scientist and inventor or, as described and conceptualized by Ursula Klein, the "hybrid expert."[22] It may be mentioned in passing that, while such hybrid experts were rather the exception among technical experts before the eighteenth century, one finds hardly any of the renowned early modern scientists – be it Johannes Kepler (1571–1630) and Galileo Galilei (1564–1642) or Isaac Newton (1643–1727) and Gottfried Wilhelm Leibniz (1646–1716) – who did not also dedicate himself to technological problems.

[18] For this literature, see, for instance, Popplow (2015).

[19] A.R. Hall (1959) 17; see also A.R. Hall (1963). Hall did not deny contacts and exchanges between learned men and practitioners in the early modern period but regarded them as fairly insignificant for the rise of the modern sciences, which he saw developing autonomously according to their own laws.

[20] A.R. Hall (1954). Hall's critique was directed against Merton (1938). In the ensuing debate, other suspects of vulgar Marxism besides Merton came into the firing line, e.g. Edgar Zilsel, Boris Hessen, or Henryk Grossmann. For the Scholar-Craftsman controversy and its background, as well as for the following, see Klein (2020) 231ff.

[21] For these experts and their significance for the new sciences, see also Long (2011).

[22] See Merton (1938) 146. For the notion "hybrid expert," see Klein (2017).

These scientists and technical experts are certainly the most important but not the only actors of an increasingly mutual openness between learned and practical knowledge in the early modern economy of knowledge, as we will see below in connection with the media of knowledge transmission and exchange.

(2) As regards *venues* where learned men or, later, scientists and practitioners could meet, establish and maintain contacts, and share mutual interests, a whole spectrum of such opportunities for encounter developed during the early modern period. At the end of the Middle Ages, flourishing cities like Ghent, Florence, or Nuremberg provided such venues in the form of artists' workshops and municipal institutions and became cultural centers alongside university cities. Beginning around 1500, a growing number of absolutist rulers sought to raise the prestige of their courts by inviting and engaging learned men and excellent practitioners competent in civil projects as well as warfare. It may suffice just to name the courts of the Montefeltro in Urbino, the Medici in Florence or Emperor Rudolph in Prague.[23]

Another venue of learned and practical knowledge was offered by sites of advanced technologies. These venues were certainly less splendid than princely courts but probably more momentous for the development of science as well as technology. Georg Agricola (1494–1555), for example, when working as a mining physician in one of the centers of silver mining, studied the technologies of mining and smelting and published his findings in a work that became the standard reference on these topics for the next 200 years. It goes without saying that he could accomplish this work only by intensive exchange with practical mining and smelting experts. Irrigation and drainage of mines as well as of marshlands was another advanced technology which attracted hybrid experts like Simon Stevin (1548–1620) or renowned scientists like Leibniz who experienced the difficulties one encounters in communication between learned and practical knowledge. Another example is Galileo's dealings with the Arsenal of Venice, which also entailed embarrassing misunderstandings.[24]

Among the workplaces that became such venues in the early modern period, the workshops of instrument makers deserve particular attention. They were meeting points of various kinds of practical mathematicians – astronomers, mathematical geographers, cartographers, surveyors, pilots and navigators – including several hybrid experts who were ideal mediators between the worlds of learned and practical knowledge.[25] Similar effects involved large practical mathematical projects like the mapping of an entire big city achieved in cooperation with various kinds of practitioners and educated men. Such cooperations worked well as "trading zones" between the professionals of these two different worlds.[26]

---

[23] For the European courts of the early modern period, see, for instance, Adamson (1998) and Elias (1983).

[24] See the chapters below on mining and mechanical engineering.

[25] For mathematical instruments, instrument makers, and practical mathematics in the early modern period, see, for instance, Damerow and Lefèvre (1985) and Bennett (1987).

[26] For the notion of "trading zones," see Long (2012).

Finally, in the seventeenth and eighteenth centuries, venues of exchange between scientific and technological expertise even assumed an institutional form, namely in the new academies of science.[27] As is well known, the first of these early modern academies, the Florentine *Accademia del Cimento* (founded in 1657), was an institution in which practical experts with or without an academic background cooperated in scientific investigations undertaken by the academy. Probably less well known is that the famous academies of science of Paris and London, also founded in the seventeenth century, as well as their later counterpart in Prussia, admitted as members experts in various practical fields without an academic training. These experts cooperated (and competed) with the learned members in assessing proposed projects for the government as well as in genuine scientific investigations.

(3) As regards *institutions of knowledge transmission*, the educational institutions organizing the transmission of learned and practical knowledge were largely separate from each other from the Middle Ages up to the eighteenth century and beyond. As a rule, learned knowledge was taught at the universities, while practical knowledge was transmitted by the system of apprenticeship and travelling journeymen. Technical experts such as architects, engineers, mining officials and so on acquired their practical knowledge and expertise principally in the same way as master craftsmen – by apprenticeship, travelling, exchange with other experts, and participation in or coordination of extraordinary projects. As we have seen above, sustained educational institutions for teaching sophisticated technological knowledge came into being only in the late seventeenth and the eighteenth centuries.

As is well known, a general school education did not exist in this period. However, as early as in the High Middle Ages, there were a few places where some schooling of practitioners was established, for example in Italy where *Abacus Schools* transmitted basic mathematical knowledge for merchants and some craft professionals. These schools were established in some cases by city authorities, mostly however by private teachers, the arithmeticians or reckoning masters (*Rechenmeister*). After the advent of printing, books or booklets provided practitioners and an interested public with mathematical or other knowledge, that is, texts that facilitated self-study.[28]

Self-study and autodidactic appropriation of learned knowledge compensated for the lack of schools and other institutions of knowledge transmission besides the universities. This must be regarded as an essential feature of knowledge transmission in the early modern period. The proficiency of engineers, architects, and particularly of practical mathematicians in mathematics and sometimes even in classic texts and theories remains inexplicable otherwise. The engineer-artist Leonardo da Vinci (1452–1519) is certainly the most prominent of these autodidacts. The more or less successful efforts of the architect and engineer Francesco di Giorgio Martini (1439–1501) or the practical mathematician Niccolò Tartaglia (1499–1557), both of

[27] For early modern academies of science, see, for instance, Feingold and Giannini (2020).

[28] See Sects. 7.2 and 7.3 of the chapter below on practical mathematics.

whom lacked university education, in translating and editing texts by Vitruvius, Euclid, and Archimedes show how ambitious Renaissance autodidacts could be.[29]

(4) As regards the *media of knowledge transmission*, the significance of media for the transmission of knowledge appears in a new light against the background that almost no school education existed in the early modern period. Yet an increasing demand for all kinds of knowledge was aroused at the turn of the fifteenth century by the developments of extended commerce, new trades and technologies, geographical discoveries, and so on. This led, on the basis of the newly invented art of printing, to a boom in practical books – such as booklets on basic mathematics for practitioners, gunners' manuals, *Books of Secrets*, *Probier-* and *Destillierbüchlein*, etc.[30]

Ordinary craftsmen are rarely found among the authors of such books but there are plenty of artists, engineer-artists, practical mathematicians, instrument makers, etc. and, particularly in the sixteenth century, physicians. Many publishers also acted as "authors," since not a few of these practical books were compilations of sections or parts of earlier books or even – according to today's standards – outright plagiarisms which was not yet generally condemned at the time. In view of the transmission of knowledge, such compilations or plagiarisms should not be dismissed; they were, like translations, no less important and sometimes even more effective transmitters of knowledge than the respective original publications.

Finally, it should not be forgotten that books – manuals, instructions, treatises – were not the only media of knowledge transmission but also drawings, plans, or diagrams. Practical knowledge could also be transmitted by presentational documents – advertisement of inventions, project proposals, or inventories. The famous early modern theaters of instruments and machines must be included here as well as their "real" counterparts, that is, arsenals and other collections of devices and/or models of devices – collections that had sometimes been established and classified especially for didactic purposes.

Summing up this section, we can state that the early modern economy of knowledge in its various dimensions brought forth a cluster of favorable conditions for interrelations between learned and practical knowledge. Moreover, when looking back from the eighteenth century to the late Middle Ages, a certain progressive development of such favorable conditions seems unmistakable. Printed texts and images succeeded manuscripts; practical booklets were followed by manuals, treatises, and textbooks on technological issues; sites of advanced technologies and, later, academies succeeded princely courts as major venues for educated men and

---

[29] By the way, most of the knowledge these men sought could not be obtained in the universities of the time.

[30] For these practical booklets, see Ferguson (1959) and Eamon (1994) as well as Sects. 3.1, 3.2 and 3.3 of Chap. 3, Sects. 4.2 and 4.4 of Chap. 4, and Sect. 7.3 of Chap. 7. – Eamon saw a connection between the broad interest in this sixteenth-century practical or how-to literature and the later efforts of systematically studying the technology employed in the arts and crafts as prominently advocated by Francis Bacon (1561–1626) and other seventeenth-century promoters of a new learning. For Bacon's idea of a "history of trade," see, for instance, Houghton (1941).

technical experts; learned knowledge was no longer the exclusive possession of traditional scholars; and university trained engineers and architects like Bernard Forest de Bélidor (1698–1761) took the place of autodidacts like Francesco Giorgio Martini or Leonardo da Vinci.

No wonder the West was (and is) proud of these developments and saw them, and particularly the Scientific Revolution, as evidence that rationality and sober empiricism, Rupert Hall's "modern scientific attitude," were the basis and guarantees of the success (and alleged superiority) of the West. The current debates about the Great Divergence pivoting around the question "Why Europe?" rightly remind us of the fact that several civilizations in history achieved comparable peaks of scientific and technological accomplishments even though, for a variety of reasons, they proved unable to stabilize and sustain this advanced level.[31] Focusing as we do in this book on science and technology alone, it is hard to see why eighteenth-century Europe could have been more successful in this respect.[32] Whatever the case, our focus is on the development of scientific and practical knowledge up to the level achieved in the eighteenth century and, more specifically, on the interrelations between these two domains of knowledge in the course of this development.

1.3 The Book's Questions and Approach

If one takes the *École polytechnique* in France or the *nützlichen Wissenschaften* pursued by the Prussian state administration in the second half of the eighteenth century as beginnings and early instances of the modern relationship between science and technology which Layton conceived of as the relationship of *mirror-image twins*, how can one account for the emergence of such early institutions? Did they result from a "scientific revolution" or were they, as we will see in the following, the fruit of a long and meandering development of interrelations and exchanges between learned and practical knowledge which can be traced back to the late Middle Ages? How can this development be adequately described and how, on the basis of such a description, can the significance of this process for the early modern history of knowledge in the West be assessed? These are the overarching *questions* this book tries to answer.

There exists a considerable amount of literature concerning several stations and events in the course of this long development process as well as its various aspects,

---

[31] For the debate about the Great Divergence, see, for instance, Pomeranz (2000) and Goldstone (2009).

[32] Among the economic and political conditions and factors that should additionally be taken into account for Europe's sustained development of natural and technological sciences, one factor deserves particular attention in my opinion, namely that industrial capitalism emerged as an economic system that compels every economic actor towards permanent innovations, forcing the state machineries to provide the necessary conditions for an adequate development of natural and technological sciences.

such as those indicated in our brief discussion of the economy of knowledge, that sustained this process. The majority of these studies focus on particular periods such as the Renaissance and particular countries like England or Italy, if not smaller localities; many of them try to locate stages of this process in an overarching cultural historical picture or dwell, particularly in recent times, on social issues such as the ambitions of practitioners who entered into commerce with educated persons or the latter's alleged or actual domination of the former.

As meritorious and indispensable as many of these studies are,[33] none of them tried to portray this process as a whole with its most essential branches. None of them made a wider comparison of the various ways in which learned and practical knowledge in the early modern age interacted in different areas of technological and scientific knowledge. What is more, many of them implicitly or explicitly took physics as a model of science, and thus highlighted mechanics and mechanical engineering as the model of all interrelations of practical and learned knowledge. By contrast, this book aims at a more complete portrait of the early modern interrelations and interactions between learned and practical knowledge. It tries to convey a new idea of the variety and disunity of these relations by discussing and comparing altogether six widely different fields of knowledge and practice.

The chosen fields are architecture, chemistry, gunnery, mechanical engineering, mining, and the various domains of practical mathematics including practical astronomy, mathematical geography and cartography, navigation, surveying and higher geodesy. Each of these fields of technologically advanced practice brought forth several kinds of technological literature which interacted in various ways with learned literature.[34] Needless to add that the chosen fields do not cover the whole spectrum of advanced practical knowledge of the early modern period; rather, they are thought to represent the *main types* of early modern technological knowledge.[35]

Mechanical engineering, to begin with, is the classic case of a field of advanced practice with a primary and well-defined counterpart in the world of learning – mechanics (statics, kinematics, and dynamics) – which had been inherited from

[33] References to many items of this literature are given in the chapters below.

[34] It is not the goal of this book to provide a comprehensive history of each single field. In the case of mechanical engineering, for example, the goal is neither to present a history of early modern mechanical engineering nor a history of mechanics; rather, the book will trace the historical development of the technological and scientific literature pertaining to this field. As to the history of early modern theoretical mechanics and early modern mechanical engineering, there is hardly any demand for another account.

[35] Two important fields of practical knowledge – agriculture and medicine – will not be considered; the latter due to limits of competence, the former, although economically the most important production sector up to the eighteenth century, is omitted as it seems dubious whether it makes sense to trace learned elements of agricultural knowledge before the eighteenth century. As regards the practical knowledge of the crafts, this knowledge is, as mentioned above (note 6), involved only implicitly insofar as crafts were part and parcel of the specialized practical knowledge of one of the chosen technologically advanced fields of practice.

Antiquity and further developed during the Middle Ages. The same holds for *Gunnery* or more precisely for its branch of external ballistics.[36]

The various domains of *Practical mathematics* must also be regarded as close relatives of this type of technological knowledge although the relations of these different domains to learned traditions (the domains of *mixed mathematics* and others) differed to some extent.

Chemistry is another type of technological knowledge. Its counterpart in the world of learning were philosophical theories of the ultimate constitution of matter. Given the metaphysical character of these theories, it remains to be investigated whether or how far exchanges with this type of learned knowledge contributed to the development of chemistry or, alternatively, whether chemistry became a modern scientific discipline independent of such exchanges and due to reflections of its practices.

Architecture and *Mining* represent yet another type of technological knowledge. These fields of knowledge, which were already regarded as "sciences" in the early modern period, might be called "conglomerate sciences" since they were technological fields that comprise a plethora of different kinds of knowledge which were united not by an overarching theory but rather by the cooperation of different professionals – craftsmen as well as various experts. Accordingly, these fields had no preferential learned counterpart but a variety of different interrelations with the world of learning.

Regarding our approach, it remains to add and to emphasize that these six fields of technologically advanced knowledge and their interrelations and interactions with learned knowledge will be investigated and discussed through a specific lens: by focusing on the technological literature. The early modern technological literature is one of our richest archives for studies of interactions of learned and practical knowledge. Hence, we will present it below in more detail than the better-known scientific literature.

References

Adamson, John. 1998. *The Princely Courts of Europe: Ritual, Politics and the Culture under the Ancien Régime 1500–1750*. London: Weidenfeld and Nicolson.

Aguillon, Louis. 1889. *L'École des mines de Paris: Notice historique*. Paris: Vve. Cn. Dunod.

Alder, Ken. 1997. *Engineering the Revolution: Arms and Enlightenment in France, 1763–1815*. Princeton: Princeton UP.

Ash, Eric H. 2019. By any Other Name: Early Modern Expertise and the Problem of Anachronism. *History and Technology* 35 (1): 3–30.

Belhoste, Bruno. 2003. *La formation d'une technocratie: l'École polytechnique et ses élèves de la Révolution au Seconde Empire*. Paris: Belin.

[36] This was the most frequently studied type of technological knowledge, and includes hydraulic engineering and shipbuilding, fields of advanced practice that are omitted here due to limits of space and time.

Bennett, James A. 1987. *The Divided Circle: A History of the Instruments for Astronomy, Navigation and Surveying*. Oxford: Phaidon Christie's.

Channell, David F. 1989. *The History of Engineering Science: An Annotated Bibliography*. New York/London: Garland Publisher.

———. 2009. The Emergence of the Engineering Sciences: A Historical Analysis. In *Handbook of the Philosophy of Technology and Engineering Sciences*, ed. Anthonie Meijers, 117–154. Amsterdam: Elsevier.

Cunningham, Andrew, and William Perry. 1993. De-Centering the 'Big Picture': "The Origins of Modern Science" and the Modern Origins of Science. *The British Journal for the History of Science* 26 (4): 407–432.

Damerow, Peter, and Wolfgang Lefèvre, eds. 1985. *George Adams: Geometrische und graphische Versuche. Nach der deutschen Ausgabe von 1795*. Darmstadt: Wissenschaftliche Buchgesellschaft.

Damerow, Peter, Jürgen Renn, Simone Rieger, and Paul Weinig. 2002. Mechanical Knowledge and Pompeian Balances. In *Homo Faber: Studies on Nature, Technology, and Science at the Time of Pompeii*, ed. Jürgen Renn and Giuseppe Castagnetti, 93–108. Rome: L'Erma di Bretschneider.

Dobbert, Eduard. 1899. Bauakademie, Gewerbeakademie und Technische Hochschule bis 1884. In *Chronik der Königlichen Technischen Hochschule zu Berlin, 1799–1899*, ed. Rektor and K.T.H. der Senat, 11–114. Berlin: Wilhelm Ernst & Sohn.

Dupré, Sven. 2017. Doing it Wrong: The Translation of Artisanal Knowledge and the Codification of Error. In *The Structures of Practical Knowledge*, ed. Matteo Valleriani, 167–188. Cham: Springer.

Eamon, William. 1994. *Science and the Secrets of Nature: Books of Secrets in Medieval and Early Modern Culture*. Princeton University Press: Princeton.

Elias, Norbert. 1983. *The Court Society [Die höfische Gesellschaft]*. New York: Pantheon Books.

Feingold, Mordechai, and Giulia Giannini, eds. 2020. *The Institutionalization of Science in Early Modern Europe*. Leiden: Brill.

Ferguson, John K. 1959. *Bibliographical Notes on Histories of Inventions and Books of Secrets*. 2 vols. London: Holland Press.

Goldstone, Jack A. 2009. *Why Europe? The Rise of the West in World History, 1500–1850*. New York: McGraw-Hill.

Hall, A. Rupert. 1954. *The Scientific Revolution 1500–1800: The Formation of the Modern Scientific Attitude*. London/New York: Longman.

Hall, Alfred Rupert. 1959. The Scholar and the Craftsman in the Scientific Revolution. In *Critical Problems in the History of Science*, ed. Marshal Clagett, 3–23. Madison: University of Wisconsin Press.

———. 1963. Merton Revisited or Science and Society in the Seventeenth Century. *History of Science* 2: 1–16.

Houghton, Walter E. 1941. The History of Trades: Its Relation to Seventeenth-Century Thoughts: As Seen in Bacon, Petty, Evelyn, and Boyle. *Journal of the History of Ideas* 2 (1): 33–60.

Kahlow, Andreas. 2000. Die ersten Jahre der Berliner Bauakademie, Vorgeschichte und Zeitbild um 1800. In *1799–1999: Von der Bauakademie zur Technischen Universität Berlin, Geschichte und Zukunft*, ed. Karl Schwarz, 32–55. Ulm: Ernst & Sohn.

Klein, Ursula. 2017. Hybrid Experts. In *The Structures of Practical Knowledge*, ed. Matteo Valleriani, 287–306. Cham: Springer.

———. 2020. *Technoscience in History: Prussia 1750–1850*. Cambridge, MA: The MIT Press.

Layton, Edwin. 1971. Mirror-Image Twins: The Communities of Science and Technology in 19th-Century America. *Technology and Culture* 12 (4): 562–580.

Lefèvre, Wolfgang. 2001. Galileo Engineer – Art and Modern Science. In *Galileo in Context*, ed. Jürgen Renn, 11–27. Cambridge: Cambridge University Press.

Long, Pamela O. 2011. *Artisan/Practitioners and the Rise of the New Sciences, 1400–1600*. Corvallis: Oregon State University Press.

———. 2012. Trading Zones: Arenas of Exchange during the Late-Medieval/Early Modern Transition to the New Empirical Sciences. *History of Technology* 31: 5–25.

Marchand, Trevor H.J., ed. 2011. *Making Knowledge. Explorations of the Indissoluble Relation between Mind, Body and Environment*. Hoboken/Chichester: Wiley-Blackwell.

Merton, Robert K. 1938. *Science, Technology, and Society in Seventeenth-Century England*. Bruges: St. Catherine Press.

Nummedal, Tara. 2007. *Alchemy and Authority in the Holy Roman Empire*. Chicago: The University of Chicago Press.

Picon, Antoine. 1992. *L'invention de l'ingènieur moderne: lÉcole des ponts et chaussées, 1747–1815*. Paris: Presses de l'Ecole nationale des ponts et chaussées.

Pomeranz, Kenneth. 2000. *The Great Divergence. China, Europe and the Making of the Modern World Economy*. Princeton: Princeton University Press.

Popplow, Marcus. 2015. Formalization and Interaction. Towards a Comprehensive History of Technology-Related Knowledge in Early Modern Europe. *Isis* 106 (4): 848–856.

Renn, Jürgen. 2020. *The Evolution of Knowledge – Rethinking Science for the Anthropocene*. Princeton/Oxford: Princeton University Press.

Roberts, Lissa L., Simon Schaffer, and Peter Dear, eds. 2007. *The Mindful Hand: Inquiry and Invention from the Late Renaissance to Early Industrialisation*. Amsterdam: Koninklijke Nederlandse Akad. van Wetenschappen.

Schatzberg, Eric. 2006. "Tehnik" Comes to America: Changing Meanings of "Technology" before 1930. *Technology and Culture* 47 (3): 486–512.

Smith, Pamela H. 2004. *The Body of the Artisan: Art and Experience in the Scientific Revolution*. Chicago: The University of Chicago Press.

Smith, Pamela H., and Benjamin Schmidt, eds. 2008. *Making Knowledge in Early Modern Europe – Practices, Objects, and Texts, 1400–1800*. Chicago: The University of Chicago Press.

[illegible], [illegible] [illegible]. [illegible] and [illegible] [illegible].

[illegible]. 19[illegible]. *[illegible]* [illegible].

[illegible]. 20[illegible]. *[illegible]* [illegible]. [illegible] University Press.

[illegible], M. 20[illegible]. [illegible] and [illegible]. *[illegible]* [illegible].

[illegible]. [illegible] [illegible].

[illegible]. [illegible] *[illegible]* [illegible]. [illegible] University Press.

[illegible]. [illegible]. *[illegible]* [illegible]. Chicago: The University of Chicago Press.

Chapter 2
Architecture

2.1 Science of Architecture

The opening sentence of Vitruvius' *De architectura libri decem*, probably the most famous treatise on architecture ever written, reads as follows: "Architecture is a science arising out of many other sciences, and adorned with much and varied learning; by the help of which a judgment is formed of those works which are the result of other arts."[1] In this definition of architecture as a science (*scientia)* the term *scientia* does not, of course, mean natural science. The science of architecture is a technological, not a natural science, that is, a science of techniques, tools, materials, and processes used and employed in the field of construction and furthermore a science of practical requirements as well as aesthetic standards that a building must meet. Like other technological sciences, the science of architecture comprises various fields of knowledge – *disciplines*, as Vitruvius put it – including elements from natural sciences concerning the various practices involved in architectural production. These fields of knowledge are interconnected not by an overarching theory but by the interaction of the practices they study, describe and explain.[2] The science of architecture is not just an umbrella of these fields but an attempt at delineating the extension and structuring the fabric of these various kinds of practical knowledge. This endeavor to present and structure the kinds of knowledge pertaining to architecture was characteristic of the architectural literature of the West from the outset,

[1] "Architectura eſt ſcientia pluribus diſciplinis, & uarijs eruditionibus ornata: cuius iuditio probantur omnia, quæab cæteris artibus perficiuntur opera." (Vitruvius: *De architectura libri decem*, book I chap. 1.) English translation by Joseph Gwilt: *The Architecture of M. Vitruvius Pollio in Ten Books* (London: Priestley and Weale, 1826, p. 3). This definition was restated by many authors up to the seventeenth century, e.g. by Giovanni Branca in 1629: "L'architettura … è una scienza di più dottrine insieme congiunte; dalla quale si approvano tutte le opera …" (*Manuale d'architettura*, book I chap. 1).

[2] See Lefèvre (2017), section I.

W. Lefèvre, *Minerva meets Vulcan: Scientific and Technological Literature – 1450-1750*, Archimedes 60, https://doi.org/10.1007/978-3-030-73085-7_2

if we take the treatise *De re aedificatoria* of Leon Battista Alberti (1404–1472), written ca. 1443, as the first comprehensive document of the Western science of architecture. That this science was able to start at once with a mature text that presented extension and structure of the field of architectural knowledge was due precisely to the fact that Alberti could build on the treatise containing the definition quoted above, that is, the work *De architectura libri decem* by the Roman architect Marcus Vitruvius Pollio (first century BCE).[3]

Vitruvius' book is the most comprehensive text on architecture (and some other technological branches) of Classical Antiquity that was transmitted to the West. As regards architecture, Vitruvius divided his work into three main sections: (1) general discussions on education and architects' competences as well as basic concepts of architecture; (2) discussions on urban construction and building materials; (3) discussions of various building types: construction of temples, construction of public and private buildings, interior construction, and hydraulic construction. Like his definition of architecture, this basic division and arrangement of the science of architecture remained the principal model for treatises on architecture in the early modern period.

To give just some examples: Alberti arranged the topics of architecture in this way: (1) Basic principles of architecture including the importance of the construction of plans, (2) building site, materials, and basic construction techniques (laying foundations, erecting walls, etc.), (3) building types, (4) decoration, (5) restoration of buildings. Two influential sixteenth century treatises – Sebastiano Serlio's (1475–1554) *Sette libri d'architettura* and Andrea Palladio's (1508–1580) *I quattro libri dell'architettura* – deviated from the model to some degree. Serlio's arrangement: (1) Geometry and perspective, (2) exemplary ancient buildings, (3) principles of architecture and the classical five orders of columns, (4) building types, (5) decoration (especially the rustic style), (6) villas; Palladio's arrangement: (1) Materials, building site, and basic construction techniques (laying foundations, erecting walls, etc.), (2) the five orders, (3) interior construction, (4) decoration, (5) categories of constructions (buildings as well as road and bridge construction), (6) exemplary ancient buildings. Two handy seventeenth-century booklets on architecture – Giovanni Branca's (1571–1645) *Manuale d'archittettura* and Georg Andreas Böckler's (1617–1687) *Compendium architecturae civilis* – stuck more closely to the model. Branca's arrangement was as follows: (1) Definition and principles of architecture, (2) materials and building site, (3) the five orders, (4) interior construction, (5) mathematics for practitioners. Böckler's arrangement was roughly similar: (1) Definition and principles of architecture (including

[3]Vitruvius' book was not forgotten during late Antiquity and the Middle Ages. About 80 medieval manuscripts of Vitruvius' *De architectura* are known. For the medieval tradition and reception of *De architectura*, see Schuler (1999). In the course of the Renaissance, the book gained the status of *the* canonical text in discourses on architecture. After the advent of printing, it was published in several editions and languages: First print edition in Latin by Giovanni Sulpicio (Rome c. 1486); several Italian editions in Latin and Italian, e.g. by Fra Giocondo (Venice 1511), Cesariano (Como 1521), Barbaro (Venice 1556 and 1567) etc.; other translations into vernaculars: into French by Jean Martin (Paris 1547), German by Walther Ryff (Nuremberg 1548), English by Henry Wotton (London 1624) etc. For these early editions and translations, see Olschki (1965) II 203 and 205, Kruft (1985) 72ff., Long (2011) 80ff.

drawing and perspective), (2) building site and materials, (3) basic construction techniques, (4) measurements of buildings, (5) the five orders.[4]

Obviously, the topics arranged in these dispositions differ greatly as regards the kinds of knowledge presupposed by each of them and, by implication, the (possible) relationship to scientific theories that can be expected from the respective kinds of knowledge. Following Vitruvius, one can and must distinguish between knowledge pertaining to the concrete execution of the construction of a building or its parts, that is, the various kinds of knowledge of the crafts involved in a construction project, on the one hand and, on the other, learned knowledge such as geometry and arithmetic or, as Vitruvius suggested, in view of urban planning, medicine and astronomy.[5] However, this does not mean that designing and planning alone, and not techniques of constructing architectural objects, are of interest in the relations between technological and scientific knowledge. In the following, therefore, we will endeavor to study the developments of early modern literature concerning constructing techniques on an equal level with those of literature on designing techniques. To this end, we will not limit ourselves to the aforementioned famous Renaissance treatises on architecture, which have shortcomings with regard to construction techniques in general and particularly regarding special engineering tasks faced by architects, such as infrastructural and hydraulic construction or military architecture.

Construction methods in the Renaissance as well as in the Baroque period did not result from principal technological changes or technological revolutions comparable to those induced by the use of iron, steel, and glass as building materials in the nineteenth century or reinforced concrete in the twentieth century. Craftsmanship in brick and natural stone masonry, along with that in carpentry and metalworking, remained the principal technical basis of the building trade. A brief look at this basis may be in order.

2.2 Crafts and Codification of Practical Knowledge

As is well known, in the early modern period craftsmen generally did not codify their practical knowledge. Masons, carpenters, and smiths acting at construction sites were no exception in this regard. They transmitted their professional knowledge to the next generation within the family or by apprenticeship and the institution of travelling journeymen. In other words, the next generation acquired the

---

[4] Alberti's *De re aedificatoria* of c. 1443 was initially disseminated as a manuscript; the first print edition appeared in 1485: *De re aedificatoria opus elegantissimum et quam maxime utile.* Florence: Laurentius; Sebastiano Serlio: *Sette libri d'architettura*, Venice and other places, 1537–1575; Andrea Palladio: *I quattro libri dell'architettura.* Venice 1570; Giovanni Branca *Manuale d'architettura.* Rome 1629; Georg Andreas Böckler: *Compendium architecturae civils.* Frankfurt 1648.

[5] *De architectura*, book I chap. 1.

knowledge of the trade not by studying books or attending schools but by oral instruction and – most importantly – by doing, since essential parts of a craft's practical knowledge cannot be acquired otherwise than by doing. The famous medieval masons' lodges acted principally in the same way. Accordingly, no literature developed on the techniques and tools of these crafts. This way of transmitting and acquiring the necessary practical knowledge of a trade worked satisfactorily when it was adapted to given local traditions and local natural conditions such as available building materials (types of stone, clay and limestone, particular wood types etc.). Under these circumstances, many building techniques were deeply ingrained and needed no explanation.

There were, nevertheless, occasions for codifying certain parts of the practical knowledge of the construction trade. A new situation arose, for example, when a local horizon was widened by attempts to adopt techniques, materials, or building types from other localities or even other civilizations. Then it became necessary to transfer information about such hitherto unknown items by representations – be it texts, or drawings, or diagrams, or measurement tables. Thus, practical knowledge of crafts involved in the construction trade did not always remain tacit, but became codified in the context of technical knowledge transfer.

The transfer of knowledge about types of masonry work for walls, floors or ceilings developed in Roman Antiquity – *opus reticulum, opus quadratum, opus incertum* etc.[6] – is a good example. Architects of the Renaissance living in Italy or other parts of the former Roman Empire could, and did, try to reconstruct these masonry techniques by studying the remains of ancient buildings. But they also found descriptions of these techniques in Vitruvius' *De architectura* (book II chap. 8).[7] Following Vitruvius' example, Renaissance authors inserted such descriptions into their own architectural treatises as well, e.g. Alberti in his *De re aedificatoria* (book III, dispersed over several chapters), Palladio in his *I quattro libri dell'architettura* (book I chap. 9–11), etc.

It must be added that the authors of these treatises generally confined themselves to providing information by text and images on ancient masonry techniques.[8] In other words, they did not discuss contemporary techniques.[9] It can generally be stated that these treatises offered almost nothing about early modern construction techniques that could possibly be the beginning of an interchange between the science of architecture and natural sciences. This holds also for their brief treatments of construction materials such as stones, sand, or

[6] See, for instance, Roth (1993).

[7] Since Vitruvius wrote his treatise in the first century BCE, it contained no information about building techniques of the Imperial Era, i.e., nothing about developed building techniques with concrete or Roman vaulting techniques. For these techniques, see Lancaster (2005).

[8] The images testify to archeological investigations.

[9] In book 8 of his *L'idea della architettura universale* (Venice 1615), Vincenzo Scamozzi (1548–1616) touched upon contemporary techniques although not in much detail. The Venetian architect Antonio Rusconi (c. 1500–1578) left behind a considerable number of engravings that depict several contemporary construction techniques in detail and certainly constitute a valuable codification of these techniques. (Rusconi created these engravings for a planned Vitruvius translation which appeared posthumously in 1590.)

bricks. It is hardly conceivable that these trivial treatments were of any value either for practitioners or for persons interested in scientific questions, who could certainly learn much more from any master mason or carpenter.

Another remarkable instance of codifications of early modern craftsmen's knowledge relates to carpentry work on construction sites. In 1561, the French architect Philibert De L'Orme (c. 1510–1570) published *Nouvelles Inventions pour bien bastir et à petits Fraiz*, a book about wooden roof constructions.[10] Even if this is seen merely as an advertising of inventions, one has to acknowledge the precise and elaborate descriptions De L'Orme provided. And, furthermore, we should note the remarkable developments in roof and ceiling constructions in the sixteenth, and particularly the seventeenth centuries.[11] These developments culminated in the eighteenth century when Niccola Zabaglia (1664–1750) published his *Castelli e ponti* in 1743, an influential tract about scaffolding.[12] In the seventeenth century, problems of roof constructions even became topics of discussions in academies.[13]

The machinery employed in the construction trade is another case in which aspects of the contemporary construction techniques were codified (mainly by drawings). This machinery was principally of the same kind as that employed in siege craft and artillery, the mining and smelting trade, or in trades using wind or waterpower. However, the intricate cranes and lifting devices employed at some construction sites were certainly something outstanding – not only for us today but for the historical actors as well. The famous machines invented by Filippo Brunelleschi (1377–1446) for the construction of the cupola of Santa Maria del Fiore in Florence aroused architects' interest long after the completion of the cathedral. That is why we have images of these machines drawn by famous figures like Leonardo da Vinci (1452–1519), Francesco di Giorgio Martini (1439–1501), or Antonio da Sangallo the Younger (1484–1546).[14] About 150 years after Brunelleschi, in 1586 Domenico Fontana (1543–1607) orchestrated the spectacular deployment of a mass of pieces of machinery for transporting and lifting to move the Vatican obelisk, and published it in 1590 with impressive illustrations (Fig. 2.1).[15]

---

[10] Philibert De L'Orme: *Nouvelles Inventions pour bien bastir et à petits Fraiz.* Paris: F. Morel, 1561. For details, see Campa (2006).

[11] In the mid-seventeenth century, Cosimo Noferi (?-c. 1663) compiled *Travagliata Architettura,* an illustrated manuscript on static problems of wood constructions – see Schlimme (2006). For roof constructions of churches in seventeenth-century Rome, see Valeriani (2006).

[12] Niccola Zabaglia: *Castelli e ponti con alcune ingegnose practiche.* Rome: N. and M. Pagliarini, 1743.

[13] For instance, in the Italian *Accademia della Vacchia.* See Schlimme (2006) and Schlimme et al. (2014) section 2.12.3.

[14] See DMD – IDs LdVCA34, LdVCA35, gm94a, gm95a, sa1449, sa1450.

[15] *Della transportatione dell'obelisco vaticano et delle fabriche di nostro signore Papa Sisto V, fatte dal cavalliere Domenico Fontana, architetto di Sua Santita.* Roma: Domenico Basa, 1590. Fontana devised and coordinated the concerted effort of 900 men, 75 horses, and countless pulleys and other machine parts.

Fig. 2.1 Lowering of the Vatican Obelisk in 1586
Domenico Fontana: *Della Trasportatione dell'Obelisco Vaticano* (1590)

Vitruvius discussed machinery, not only cranes and suchlike but also water lifting devices, water clocks, and other devices in Book 10 of his *De architectura*. Accordingly, one finds descriptions and depictions of machines in early modern editions or translations of Vitruvius, e.g. in Cesare Cesariano's Italian edition of 1521 or in Daniele Barbaro's annotated Italian edition of 1556 and 1567. Machinery is also dealt with in other early modern treatises on architecture, e.g. in Antonio Averlino's (Filarete – 1400-1469)) *Trattato di architettura* of c. 1462, and particularly in treatises on military architecture, e.g. Buonaiuto Lorini's (1545–1611) *Delle fortificationi* of 1597 and, most elaborately, Francesco di Giorgio Martini's *Trattati d'architettura* of c. 1490. As is well known, Francesco was as much a mechanical engineer as an architect.[16] (The chapter on Mechanical Engineering below discusses early modern machinery in detail.)

Finally, there is a considerable amount of codification of the practical knowledge of late medieval as well as early modern stone masons, particularly concerning geometrical techniques applied in stone cutting. Since this field of practical knowledge overlaps with architectural design techniques, we will discuss it below when dealing with these techniques.

2.3 Design I – Buildings Types and Their Decoration

The question of how a sacred building, a castle, or a city hall should be designed was, and still is, a matter of orientation to traditions and exemplary realizations (leaving aside economic considerations). The architect or master builder in charge of a concrete construction project thus had to be knowledgeable about how the building type at hand had been traditionally designed and also had to be familiar with recent exemplary realizations. He acquired this knowledge by studying existent architecture, travelling around and recording his experiences in logbooks. The emergence of the Renaissance treatises on architecture meant that he could also resort to codified knowledge on this subject.

Two points must be added here: First, these treatises on architecture – regardless of whether printed books or manuscripts like Alberti's treatise or the manuscripts of Vitruvius' *Decem libri* before the end of the fifteenth century – did not displace personal notebooks with records of exemplary architecture. Copies of such records, particularly pictorial records, circulated incessantly among professionals as well as clients and interested individuals during the early modern period.[17] The possible significance of this unpublished literature for an interchange of architectural and scientific knowledge is still an unexplored question. Second, the knowledge of traditional or exemplary architecture was not, and could not be, codified by means of words alone. Graphical representations – drawings or plans – were indispensable means of communicating this architectural knowledge. Consequently, drawings, plans, and diagrams should be regarded as technological "literature" on the same level as texts.

---

[16] Cesariano, Cesare: *Di Lucio Vitruvio Pollione de architectura libri decem.* Como: Gotardus de Ponte, 1521. Daniele Barabaro: *I dieci libri dell' architettura di M. Vitruvio.* Venice: Franceschi and Chrieger, 1556. Antonio Averlino: *Trattato di architettura*, c. 1462 (original not extant; manuscript copies, e.g. Biblioteca Nazionale, Florence: Codex Magliabechianus II, I, 140). – Buonaiuto Lorini: *Delle fortificationi libri cinque*. Venice: G.A. Rampazetto, 1597. Francesco di Giorgio Martini's *Trattati di architettura ingegneria e arte* militare, c. 1490, was disseminated by manuscript copies, e.g. Codice S (Siena, Bibl. comunale, cod. S.IV.4) and codice M (Firenze, Bibl. nazionale, Magliabechiano II.I.141, parte 1). Modern edition: Francesco di Giorgio Martini: *Trattati di architettura ingegneria e arte militare.* Corrado Maltese (ed.), 2 vols., Milan: Edizioni di Polifilo 1967.

[17] The earliest known notebook is the logbook of Villard de Honnecourt, an architect active in the thirteenth century. Famous Italian notebooks of the time around 1500 and collections of such drawings are Giuliano da Sangallo's *Taccuino Senese* (Biblioteca Communale, Siena, S.IV.8), the *Codex Mellon* (Pierpont Morgan Library, New York, 1978.44.88) or the *Codex Coner* (Sir John Soane's Museum, London). See Merrill (2017) and Brothers (2017).

Whereas master builders of the Middle Ages tried to be on a par with the state of art in contemporary architecture, Renaissance architects had to be knowledgeable about the main features of ancient Roman architecture as well. That is why almost every Renaissance treatise on architecture contains a section on the five orders of columns (Tuscan, Doric, Ionic, Corinthian, and Composite) along with their bases, capitals, architraves etc. and sometimes also a section about ancient building types. The knowledge these treatises provided about ancient architecture did not stem from Vitruvius or other literary sources alone but also from archaeological investigations carried out by architects. At this point we encounter a commerce between architectural and scholarly antiquarian knowledge aside from archeology, including humanistic philology.[18]

It goes without saying that the invention of print and the dissemination of ideas and information by printed books or booklets was a major advance with significant consequences for the spread of architectural knowledge, just as it was for knowledge in other technological branches. Equally important for the dissemination of architectural knowledge was the invention of mechanically reproducing images by means of woodcuts or engravings/etchings and the printing press. This was particularly important because of the role of graphical representations in the transmission of knowledge in this technological field, which could be compared with the role of such representations in botany and other branches of natural history. On the basis of printed images and plans, some of the sixteenth-century treatises on architecture assumed the shape of pattern books transmitting knowledge on architectural elements – building types, parts of buildings such as portals or staircases, and items of the building's decoration – mainly by images and plans of these elements along with explanations.

This holds not only for Serlio's *Sette libri* but to some extent also for Palladio's *Quattro libri*. The book on the five orders of Giacomo Barozzi da Vignola (1507–1573) contains only engravings with clarifications.[19] It was due not least to their pattern book character that these particular treatises gained a remarkable impact on the architectural design of the early modern period and were translated and reissued again and again up to the beginning of the eighteenth century.[20]

[18] The fact that Vitruvius' *De architectura* were transmitted to the West without any of its accompanying drawings caused serious problems as to how the text should be understood. All drawings and diagrams in Vitruvius editions of the sixteenth century must, therefore, be taken as interpretations – interpretations of the text based not only on philological scholarship but also on serious archeological investigations. To overcome the problems humanists as well as architects experienced with the Vitruvian text, a special team of linguistic experts was convened when the Sienese engineer and architect Francesco di Giorgio Martini served as chief architect at the court of Urbino. The results of this commission were, however, rather disappointing. See Olschki (1965), vol. I 129.

[19] Giacomo Barozzi da Vignola: *Regola delle cinque ordini d'architettura*. Rome: unknown publisher, 1562.

[20] In the seventeenth century lavishly illustrated books on contemporary exemplary buildings of Baroque Rome, e.g. by Valerien Regnart, Giovanni Giacomo Rossi, Domenico De Rossi, and Guarino Guarini can be regarded as counterparts to books on exemplary buildings of ancient Rome published in the sixteenth century.

There is scant evidence of possible exchanges between this literature on design types and sciences other than the 'antiquarian science'; literature on design techniques seems much more promising in this regard.

2.4 Design II – Measures and Rules

Whatever type of building is to be designed, it must be adapted to the intended building site. The design process thus presupposes a sufficiently exact measurement of this ground. In certain cases, particularly when fortresses were to be constructed, this measurement could even directly involve landscape surveying, including levelling. The same holds for city planning, hydraulic engineering, or road construction. That is why architects engaged in such projects had to be conversant to some degree with the methods of surveying, or at least able to call on and cooperate with professional surveyors. This became a real challenge in the early modern period since, as is well known, the art of surveying made significant advances through figures like Gemma Frisius (1508–1555) and Leonard Digges (c. 1515 – c. 1559) in the sixteenth century and beyond. As we will deal below with early modern surveying in the chapter on Practical Mathematics, it suffices to remark here only that architects, and in particular those with engineering tasks, should be regarded as mathematical practitioners in a way.[21] This will become increasingly clear in the following.

The recruitment of architects with surveying competence could actually become a serious problem in principalities or city states where the demand for performance of such engineering tasks increased. Take, for example, the dukedom of Milan in the sixteenth century with its extension of hydraulic projects in the Po valley. This led to the foundation of the first educational institution for architects in the West, the *Collegio degli Architetti, Ingegneri e Agrimensori* founded in 1563.[22]

This indicates a development towards specialization among architects – a development that anticipated the later differentiation between architects and construction engineers and which became manifest at the time particularly in the differentiation between civil and military architecture. Here we shall leave aside the vast early modern literature on fortification which was booming at the time because of a sequence of profound changes of the *manieres de fortification* in consequence of artillery improvements.[23] This represents a co-evolution between the science of architecture and the contemporaneous science of siege and artillery which we shall not explore further.[24] For the purpose of our investigation, however, we should stress that this development of military architecture considerably enlarged the range of geometrical competences required of a military architect.[25]

A major task of designing a building was, and still is, determining its measurements and proportions, that is, the ratio of its elements such as the ratio of the width

[21] See Gerbino and Johnston (2009).

[22] See Scotti (1983).

[23] See Duffy (1979/1985), Arnold (2002), and Bürger (2013). See also Kruft (1985) section 9.

[24] The case of ballistics will be discussed in the chapter on Gunnery in the present volume.

[25] See Marten et al. (2012).

to the height of a room, a church nave, a portal, and so on. With regard to such proportions in the case of the different orders of columns, Vitruvius explained a method of deriving the measurements of all the parts from the length of a "module" (usually the radius at the basis of the respective column); that is to say, the measure of each of the parts was determined as a multiple or a fraction of the "module." Similar module systems occur consequently in each of the Renaissance treatises on architecture and became a model for determining the measurements and proportions for all kinds of structural elements. In principle, this module method does not presuppose any highly sophisticated arithmetical competences on the part of the architect.[26]

Less inventive architects could make use of the patterns and plans in some of the architectural treatises, e.g. those of Serlio or Palladio, which offered measurements and proportions. But Renaissance architects also knew and tried out a more ambitious system of proportions, namely one based on the proportions of the (ideal) human body. Again, the model for this was found in Vitruvius and visualized by architects and artists like Francesco di Giorgio Martini, Cesare Cesariano and, most famously, Leonardo da Vinci (Fig. 2.2).

However, rules for determining the proportions between the parts or elements of buildings do not always refer back to the design methods of Antiquity transmitted by Vitruvius. Similar rules and methods were also in use in the West independently of this scholarly tradition as the so called *Werkmeisterbücher* show which appeared in Germany at the turn of the fifteenth century.

Werkmeisterbücher were "textbooks" or – to put it less anachronistically – didactic booklets written by master builders for prospective master builders. To date, only six such booklets are known.[27] Three of them[28] deal with geometric constructions for designing special elements of Gothic sacred buildings (pinnacles and ornamental gables with tracery). The three others are far more comprehensive as regards the construction of Gothic sacred buildings; the instructions they give include those concerning the dimensions and proportions of several types of churches, transmitting practical rules for the implementation of the planned building sections on the building site. They do not, however, deal with profane buildings. Moreover, they focus on the design process, not on technical questions such as construction techniques, materials, tools or machines employed in the construction process.[29]

---

[26] However, calculations of measures had to be done in units of length that differed from one country or city state to another and demanded conversions that were far from trivial. See, for instance, Schlimme et al. (2014) section 2.6.1.

[27] Three of them appeared in print: Matthäus Roritzer: *Büchlein von der Fialen Gerechtigkeit* (Regensburg, by the author) 1486; idem: *Geometria Deutsch* (Regensburg, by the author) 1487/88; Hans Schmuttermayer: *Fialenbüchlein* (Nuremberg: Georg Stuchs) 1489. Of the other three booklets, only manuscript copies or copies of copies are known: Anonymous: n.t. (*Wiener Werkmeisterbuch*) (Albertina Wien Cim VI 55) after 1500; Anonymous: *Von des Chores Maß und Gerechtigkeit* (original no longer extant) after 1500; Lorenz Lechler: n.t. (*Unterweisungen*) (original no longer extant) 1516. A diplomatic transcription of all of these books is provided by Coenen (1990).

[28] The booklets by Roritzer and Schmuttermayer.

[29] For a thorough analysis of these *Werkmeisterbücher*, see Coenen (1990). For further literature, see https://de.wikipedia.org/wiki/Werkmeisterb%C3%BCcherr

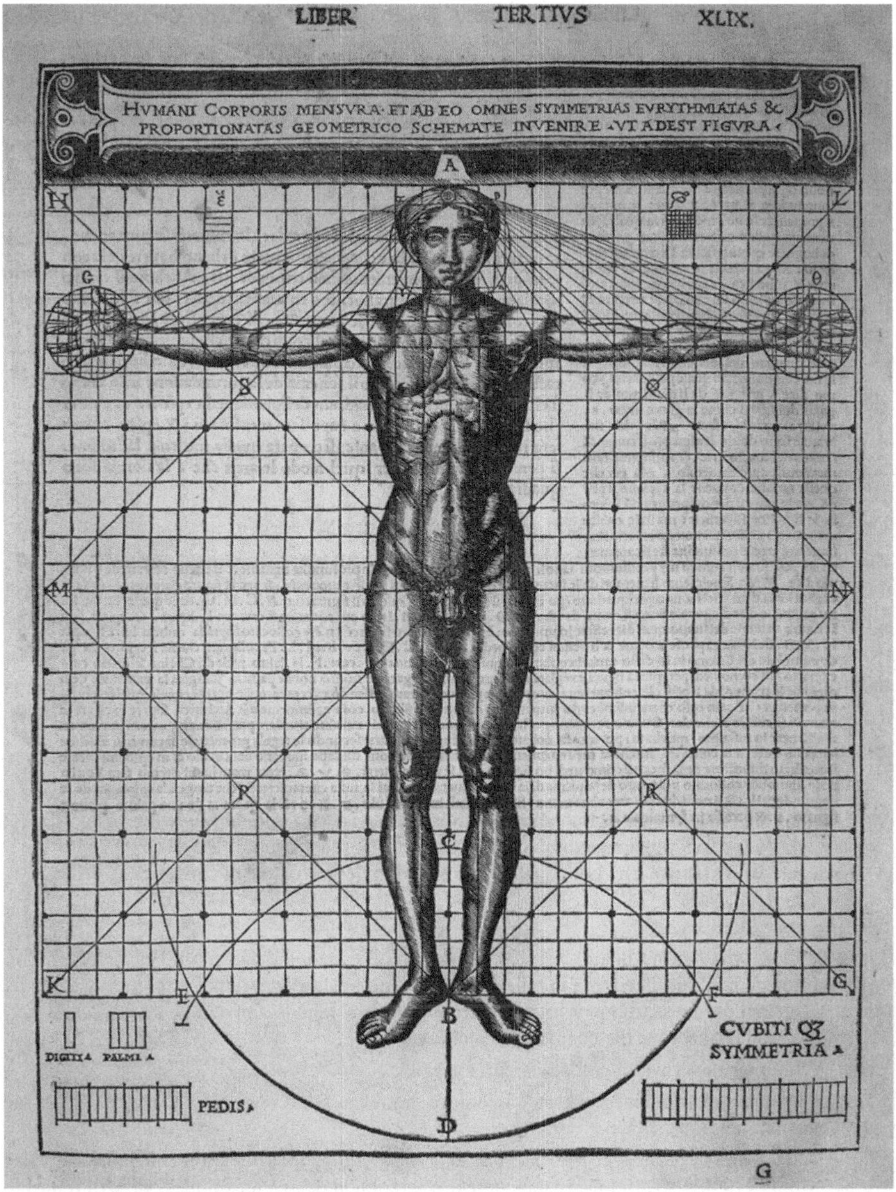

Fig. 2.2 Cesare Cesariano's Vitruvian Man (1521)
Cesare Cesariano (1969) III, p. xlix

These booklets must be regarded as true originals. There was no literary model or tradition their authors could possibly have followed.[30] Moreover, as they were written by master builders[31] and not by humanists like Alberti, these booklets appear clumsy and awkward. Thus, it might seem inappropriate to count them among Renaissance treatises on architecture. However, in one respect it is not far-fetched to do so because they, or more precisely, three of them,[32] principally teach the development of dimensions and proportions of sacred buildings in the same way as the treatises in the tradition of Vitruvius do for the classical building types. They do this by taking one dimension – in this case, the clear span of the church choir – as the basic dimension from which all others can be derived by means of elementary arithmetic.[33]

2.5 Design III – Perspective Drawings and Orthogonal Plans

Drawing is the means by which measurements and proportions get integrated and transformed into the design of a building. The Italian term *disegno*[34] means "design" as well as "drawing" and can be taken to indicate that drawing became the chief technique and design medium in the early modern period.[35] In contrast to illustrations in other technological fields of the time, such as mechanical engineering or mining,[36] those in architecture were created in drawing styles or drawing languages that were based on geometrical constructions, namely drawings in (several sorts of) perspective and orthogonal plans.

It seems that medieval master builders did not use drawings for designing a sacred or profane building.[37] This was possibly due mainly to the lack of suitable drawing materials, particularly the lack of paper, which first became an affordable item in the fifteenth century. For it is clear that the master builders had command of the technique of constructing orthog-

[30] It is not clear whether or not another *Geometria Deutsch* written by a certain Hans Hoesch preceded Roritzer's booklet. There is a parallel to these *Werkmeisterbücher*, that is, to texts written by practitioners for fellow practitioners, namely German gunners' manuscripts (*Büchsenmeisterbücher*) of the late Middle Ages; see the chapter on Gunnery, Sect. 3.2.

[31] Hans Schmuttermayer was a goldsmith and engraver.

[32] The two anonymous manuscripts and Lechler's booklet. For Lechler's *Unterweisungen*, see Shelby and Mark (1979).

[33] For this comparison of *Werkmeisterbüchern* and Renaissance treatises on architecture, see also Werner Müller (1990) 291. Below, in the context of graphical design techniques, we will return to the geometry taught in the booklets by Roritzer and Schmuttermayer.

[34] As is well known, "disegno" was a key concept of art theories of the Italian Renaissance. For the philosophical background of this concept, see, for instance, Panofsky (1924) 29ff., Wolfgang Kemp (1974), and Barzman (2000) 145ff.

[35] For the following, see Lefèvre (forthcoming).

[36] For the pictorial language of late medieval and early modern machine drawings, see McGee (2004); for the pictorial styles used in early modern literature on mining, see Lefèvre (2010).

[37] See, for example, Booz (1956).

onal plans, as is shown by such plans on walls or floors of medieval cathedrals (*Ritzzeichnungen*) as well as by the splendid elevations traced on parchment (*Planrisse*).[38]

Both techniques, that of constructing plans as well as drawings in perspective, were known and practiced in Antiquity. The former seems to have survived during late Antiquity and the early Middle Ages. Aside from a reference to some Roman murals, an ancient practice of drawing in perspective was indicated by only one term – scenography (*scaenographia*) – in Vitruvius (book I chap. 2). Thus, this technique needed to be re-invented in the Renaissance.

In this context it is interesting that the re-invention of rendering in perspective was not, as one might presume, first accomplished by painters from whom architects then learned it. Rather, this re-invention was no less due to figures like Brunelleschi or Alberti who excelled in architecture than to painters like Massaccio (1401–1428) or Piero della Francesca (1415–1492). This also holds for the early Renaissance literature that dealt with, explained, and taught techniques of constructing renderings in perspective: Alberti's discussion of perspective in his *Della pittura* of 1435 preceded Piero della Francesca's *De prospectiva pingendi* of c. 1480 and Albrecht Dürer's (1471–1528) *Underweysung der Messung* of 1525.[39]

Alberti's text focused mainly on the optical aspect of perspective, and that of Piero's and Dürer's treatises on the geometrical construction of drawings in perspective. However, these treatises contain no mathematical discussions of the geometrical presuppositions of perspective drawings. Rather, they are step by step instructions in the method of knowledge transmission characteristic of the sphere of crafts. Both authors were well conversant with Euclidian geometry, Piero outstandingly so, but obviously more interested in Euclid's geometrical constructions than in his theoretical deductions and proofs. Both treatises nevertheless received attention from historians of mathematics, who regarded them as the beginning of a development that led to a new, non-Euclidean mathematical discipline, namely projective geometry.[40]

In his *De re aedificatoria* (book II chap. 1), Alberti assigned the practice of drawing in perspective to the art of painting and stressed that architects should represent architectural subjects not in perspective, but by means of orthogonal plans – ground plans, elevations, and sections. In doing this Alberti demarcated drawings that do not render true angles and distances (that is, drawings in perspective) from those that give correct measurements (that is, orthogonal plans).[41] However, as Italian architectural drawings from the second half of the fifteenth and the first decades of

[38] For *Ritzzeichnungen*, see, for instance, Schoeller (1989) or Pacey (2007) chap. 2; for *Planrisse*, see, for instance, Böker (2005), Köpf (1977), Recht (1989).

[39] Leon Battista Alberti's *Della pittura* and Piero della Francesca's *De prospectiva pingendi* were disseminated only in manuscripts form in the fifteenth and early sixteenth centuries. Albrecht Dürer: *Vnderweysung der Messung mit dem Zirckel vnd Richtscheyt*. Nuremberg: Hieronymus Andreae 1525. For perspective rendering in the Renaissance, see Panofsky (1997), Martin Kemp (1992) part I, and Andersen (2007).

[40] See Andersen (2007); see also Judith Field in an article on Desargues published in 1995 on a website of the University of St. Andrews School of Mathematics (http://www-history.mcs.st-and.ac.uk/Biographies/Desargues.html)

[41] For this demarcation, see also Thoenes (1993).

the sixteenth century unmistakably show,[42] architects had become equally accomplished in perspective draftsmanship and plan construction. In this period, the rendering of building interiors in perspective seems to have become standard practice among architects and often replaced the wooden models which, at least in Italy, had been an indispensable medium of visualization in the consultation processes between client and architect.[43]

However, Alberti's advice and admonition to represent architectural subjects by means of orthogonal projections was by no means lost on the architects of the age. Besides drawings in perspective, they traced orthogonal plans as well as further kinds of geometrical diagrams such as stereometric layouts, templates, and so on, that is, several kinds of working drawings needed by craftsmen in the concrete construction process. The language of orthogonal plans and that of renderings in perspective were both indispensable languages in architecture. The latter can in no way be regarded as more "advanced" than the former, let alone as a graphical language that replaced the former. Both were needed in the design process as well as in the communication processes with the client, and furthermore in publications for a wider public. But only exact orthogonal plans constituted the prerequisite condition for constructing the aforementioned types of working drawings.

Although plan construction was an age-old drawing technique and in an advanced stage of development at the end of the Middle Ages, a profound innovation of this technique was invented and introduced around 1500, namely the combined views technique.[44] This technique consisted basically of a graphical procedure by which two plans of an object – say the ground plan and one elevation (both of the same scale) – are arranged in relation to one another so that a third plan – say, an elevation perpendicular to the first one, can be derived just by drawing out lines given in the initial two plans (Fig. 2.3). This technique developed into the standard procedure in architectural plan construction and is still applied today (with or without CAD). From the perspective of the history of mathematics, the procedure of combined views is nothing but descriptive geometry in its first manifest stage of development.[45] It can be attributed to Albrecht Dürer and Antonio de Sangallo the Younger. It should be noted, however, that the two artists could, and did, refer to Gothic design techniques, although initially not to those for tracing *Planrisse* but those used by stone dressers.[46] The creation of shapes imparted to stones by stone masons belongs to design as well as construction and thus constitutes a link between these

---

[42] See, for instance, the notebooks and portfolios mentioned above (note 17); see also Huppert (2015).

[43] See Lepik (1994); see also Alberti *De re aedeficatoria* book II, chap. 1.

[44] See Lefèvre (2004).

[45] See Camerota (2004).

[46] Ibid. As for Sangallo, who published no treatises, his handmade architectural drawings show that he had full command of the combined views technique. Dürer did not present any geometrical discussions but just instructions on how to derive plans from plans in his writings, particularly in the *Underweysung der Messung* of 1525 but also in his *Vier Bücher von menschlicher Proportion* of 1528.

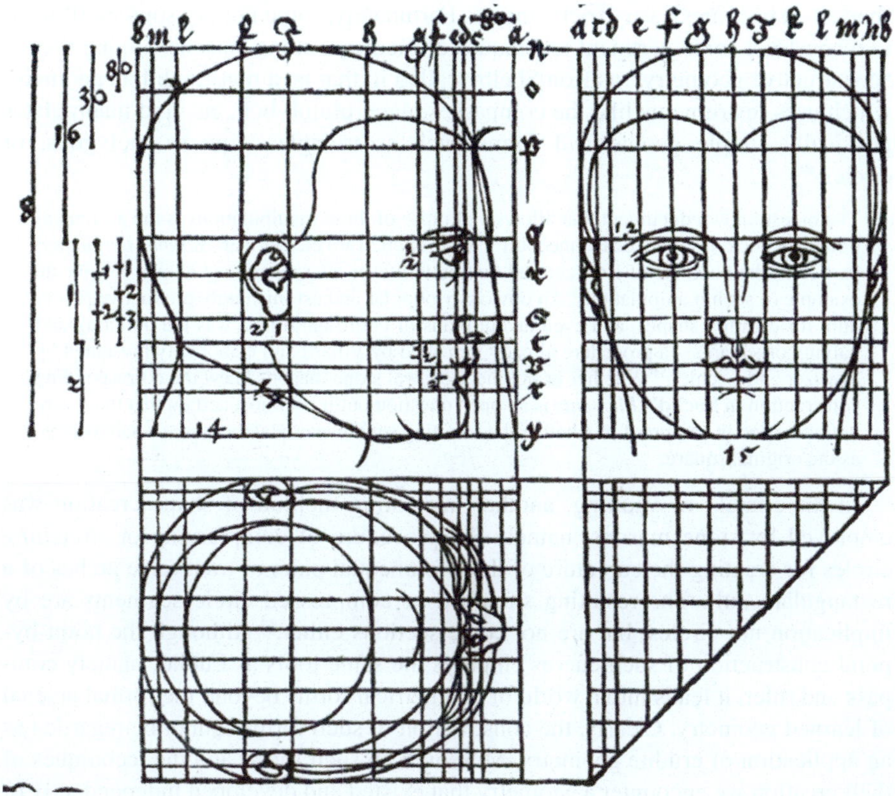

Fig. 2.3 Combined Orthogonal Projections (1528)
Albrecht Dürer (1969) I, f. E2v

branches of architectural practice. That is why we shall next consider several aspects of stone masonry.

2.6 Constructive Geometry, Stereotomy and Descriptive Geometry

Stone masonry[47] as employed in the construction of Gothic cathedrals requires the exact anticipation of bearing joints of ashlars, that is, the shapes of stone surfaces that are thought to be adjacent to one another. This is a more or less trivial task if it only involves creating rectangular shapes.[48] In all other cases – oblique or curved

[47] For the following, see Lefèvre (2017) section 3.

[48] This may have been an important task when prefab natural stones were ordered from distant quarries, a logistical practice that became more and more customary from the thirteenth century on. See Kimpel (1983), Werner Müller (1990) 126ff., Hurx (2018) chap. 3–7.

shapes – it becomes a very tricky matter. Fortunately, some literary sources allow us to reconstruct some pertinent techniques of medieval stone masons, among them a "constructive geometry," as Ron Shelby called it, that is, a non-Euclidian geometry which uses instruments like the compass, square, plumb bob, etc. and manipulates physically simple geometrical forms such as triangles, squares, polygons, or circles.[49]

The oldest known document that allows a glimpse of these manipulations is the aforementioned logbook of Villard de Honnecourt.[50] More explicit are the German *Werkmeisterbücher*, particular the booklets of Roritzer and Schmuttermayer also mentioned above. To give an example of such manipulations: An obviously popular and astonishingly productive procedure for creating shapes, and even layout plans of entire buildings, was the doubling and halving of squares – a procedure already indicated by Villard and elaborately presented by Roritzer and Lechler.[51] Whether or not the medieval stone masons knew the corresponding construction in Euclid's *Elements* is an open question but it is almost certain that they were not interested in any proof of whether the resulting square was exactly twice or half as large as the original square.

In the context of vaulting, another interesting method of shape creation was employed by stone masons, namely the procedure of compressing or stretching circles for creating the curvature of the two unequal pairs of transverse arches of a rectangular vault. The resulting stretched or compressed circle segments are by implication not circles, but are not conic sections either.[52] Although the point-by-point construction of such curves employs the same tools as Euclid, namely compass and ruler, it leads into a world of geometrical forms beyond the formal arsenal of learned geometry. Clearly, the construction of such forms cannot be regarded as an application of erudite geometry. Moreover, in such forms and the techniques of their creation we encounter a geometry that existed and developed independently of erudite geometry; that is to say, we encounter practical geometry as a second realm of geometry besides theoretical geometry.

This procedure of deforming circles was known in Antiquity and employed for creating the curvature (*entasis*) of columns, as Lothar Haselberger discovered at the Apollo temple in Didyma.[53] In the Renaissance, several methods of *entasis* construction were invented and published, for instance with the aforementioned engravings of Vignola's *Regola delle cinque ordini d'architettura*.[54]

Another stone mason's procedure deserves particular attention. It concerns the curvature of the stones which form the rib of a groined vault when jointed together.

---

[49] Shelby (1972). See also Bork (2011) and Lefèvre (2017) section 3.

[50] See folios 39 to 41 of this thirteenth-century logbook. Modern editions: Hahnloser (1972) or Barnes (2007).

[51] See Hahnloser (1972), "Abbildungen" 91 and 99, and Shelby and Mark (1979); for *Werkmeisterbücher*, see note 27 above.

[52] See fig. 33 in Albrecht Dürer's *Underweysung der Messung* (book I).

[53] See Haselberger (1999).

[54] See Becchi (2014a).

(The alignment of the stones' profile was achieved by templates.) The stone masons found this curvature by a projection procedure called *Bogenaustragung* (the "Dresden Method" of vault projection) – (Fig. 2.4). By this method, the horizontal extension of the rib segment as given in the ground plan of the vault was projected onto the vertical plan of the rib given in the elevation of the vault.[55]

Projection methods of medieval stone masons like this deserve particular attention because they evolved into design methods of utmost sophistication employed in the stone masonry of the Renaissance.[56] This architectural technique flourished particularly in sixteenth-century France where intricately shaped elements of profane buildings – staircases, entire vaults, vaulted corridors, oriels like the famous trompe

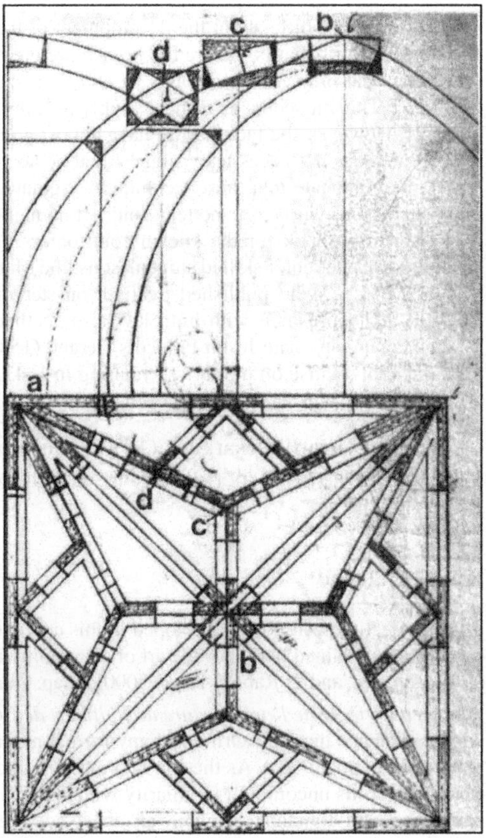

Fig. 2.4 Vault Projection "Dresden Method" (*Bogenaustragung*) J. Fracht von Andernach's *Musterbuch* (sixteenth c.) – Müller (1990), p. 178

[55] The similarity of this method to the combined views method discussed above can hardly be overlooked. For *Bogenaustragung*, see Müller (1990) chap. 5.4.1 and 5.4.2.

[56] For the principal difference between the Gothic and Renaissance mode of using natural stones, which I cannot explore here, see Rabasa-Díaz and Calvo-López (2009), particularly section 3.

of the *Chateau d'Anet* – were built out of dressed stones.[57] In the context of our investigation, it must be stressed that these sophisticated techniques of geometrical projection cannot be regarded as applied mathematics. There was no existent theory of geometrical projection, no descriptive geometry that could be applied. Rather, these techniques represent a highly developed practical geometry that existed independently of and alongside the erudite geometry of the age. It is, on the other hand, hardly imaginable that artists and master masons could have developed these techniques without considerable geometrical knowledge. Thus, one can probably say these practitioners applied geometrical theories in the tradition of Euclid – however, and this is the decisive point here, they did so not in order to map solids onto planes, but to invent and refine their own techniques of such mapping.

We know very little about the development of these techniques in the first half of the sixteenth century, that is, before the French architect Philibert De L'Orme published plans of geometrical projections in books III and IV of his *L'Architecture*, which first appeared in 1567.[58] Thus, it seems as if these highly accomplished mapping techniques surfaced suddenly. However, the fact that we have almost no documents of the development before De L'Orme does not necessarily suggest that he accomplished a stroke of genius. Rather, this fact may indicate that these techniques originated and reached maturity in the practice of stone masons who wrote no texts and left no marks besides the works they built.[59]

When we look for the known contributors to the further development of these techniques we encounter skilled stonemasons like Mathurin Jousse (1607–1650), master mason of La Flèche, who published treatises on stereometric constructions. And we encounter more and more men with high levels of mathematical competence, and even scholarly mathematicians. The Jesuit François Derand (1588–1644), for example, who published an influential treatise on mortarless vaulting in 1643, taught mathematics at the Jesuit college at La Flèche and served as one of the order's architects.[60]

In 1640, Girard Desargues (1591–1661) published a four-page draft about geometrical projections pertaining to advanced stereotomy that addressed fellow

[57] The projection techniques developed in this context of advanced stereotomy are far too complicated to be explained here. For the art of stereotomy in the long sixteenth century see Sakarovitch (1998), chap. 1 and 2; Rabasa-Díaz (2000), chap. 2 and 3; Camerota (2004), section 5.

[58] *Le premier tome de l'architecture de Philibert de l'Orme*. Paris: Federic Morel1567. There is no known evidence that de l'Orme had any formal mathematical education (see Blunt (1958), chap. 1; see also Potié (1996)). As the son of a master builder who became a master builder himself, he stood out for his uncommon familiarity with classical architecture acquired through journeys and commerce with humanists. He may also have acquired his mathematical knowledge through such commerce.

[59] "As proven by the names [e.g. *trompe de Montpellier, trompe quarrée* etc.] still used to define some of these vaults, they are architectural types derived from the Romanesque and Gothic tradition of Southern France." Camerota (2004), 203.

[60] Mathurin Jousse: *Le secret d'architecture découvrant fidèlement les traits géométriques, coupes, et dérobemens nécessaires dans les bastiments: enrichi d'un grand nombre de figures, adioustées sur châque disours pour l'explication d'iceux.* La Flèche: Georges Griveau, 1642. François Derand: *L'Architecture de voûtes ou l'Art des traits et coupe des voûtes.* Paris: chez Sebastian Cramoisy, 1643.

mathematicians rather than stonemasons.[61] Historians of mathematics regard this draft as the birth certificate of descriptive geometry, that is, "a new, non-Greek way of doing geometry," as Judith Field put it.[62] Desargues' draft is also of particular interest in another respect, namely as a document that marks the point where a theoretical descriptive geometry detached itself from the practical geometry of the stonemasons.[63] It took another century and a half before this new, non-Greek geometry assumed its classical theoretical form in Gaspard Monge's *Géométrie descriptive*, published shortly before 1800.[64]

2.7 Construction and Statics

There is almost no literature on construction subjects written by master builders or architects of the Middle Ages and the early modern period. The *Werkmeisterbücher* and Philibert De l'Orme's treatises on architecture and wood construction constitute the most explicit literature known from the time before 1600. This comes as no surprise as regards standard construction tasks such as erecting walls or covering roofs, since the required knowledge for accomplishing such tasks was transmitted by apprenticeship and not by books. Even the knowledge required for more challenging tasks such as constructing arches or vaults appears to have been transmitted in this way. The knowledge of constructing several types of vaults – barrel vaults, so-called monastery vaults, and so on – and even of constructing spherical cupolas was obviously never lost during late Antiquity and the Middle Ages despite the fact

---

[61] Gérard Desargues: *Brouillon project d'exemple d'une manière universelle du S.G.D.L. touchant la practique du trait à preuves pour la coupe des pierres en l'architecture.* Paris: published by the author 1640.

[62] In the 1995 article on Desargues mentioned above (note 40), she refers to projective geometry. But she could have said the same about descriptive geometry, since the stories of these two disciplines are strikingly parallel. "While stereotomy, together with carpentry, provides one of the richest examples of the uses of applied geometry, it is also at the root of a branch of erudite geometry, namely descriptive geometry. To sum up the situation, one might say that stereotomy is to descriptive geometry what perspective is to projective geometry. The parallel between the evolution of stereotomy and perspective is indeed striking. Both practices developed during the Gothic period – whether on stone cutting work sites or in painters' workshops. The first treatises were edited during the Renaissance and the mathematicians of the 'Monge School' explicitly theorized stereotomy and perspective at the end of the 18th and beginning of the nineteenth century" (Sakarovitch (2003) 72). It remains to add that stereotomic techniques did not originate with Gothic architecture but can be traced back to times before Classical Antiquity; see, for instance, Semper (1863), 10. Hauptstück.

[63] However, at the request of Desargues the engraver Abraham Bosse (1604–1670) published a book on stereotomy thought to be more comprehensible to stone masons: *La pratique du trait à preuves de M. des Argues Lyonnois pour la coupe des pierres en Architecture*. Paris: Claude Jombert, 1643.

[64] Monge [1798/99]. See also Dhombres (1992).

that no codification of this knowledge is extant if it ever existed at all.[65] As we have seen, only knowledge about geometrical techniques in connection with stone dressing was codified at the end of the Middle Ages.

This negative finding has always been regarded as strange and particularly unsatisfactory in view of the sophisticated and daring constructions exhibited by cathedrals of the High Gothic period. Since the displacement of romantic projections into this architecture in the mid-nineteenth century due to the interest in its construction by a sober engineer, Eugène Violett-le-Duc (1814–1879), up to the present day a host of conjectures have been proposed about what the medieval master builders may have aimed to achieve with various constructional details of these cathedrals. And alongside the romantic fascination with Gothic architecture, which was pushed back but not replaced, speculations bloomed about the secrets of the medieval masons' lodges. In view of our theme, namely an actual or possible commerce between technological and scientific literature in the realm of early modern architecture, both kinds of speculations appear irrelevant.[66]

This negative finding means in particular that no literature on static aspects of constructions was produced before 1600. The stability of buildings – Vitruvius' *firmitas* – was, of course, a major concern of medieval and early modern master builders and architects. They were too well aware of building collapses such as that of the cathedral of Beauvais in 1284 to be careless in this respect. It seems, however, that the master builders of the age generally relied on well-established ways of constructing precarious parts of a building, e.g. the system of piers and buttresses of a Gothic cathedral, and adapted them to the building at hand according to certain rules.[67] Since the master builders of the age did not leave texts (or diagrams) on issues of statics, we are reliant on a few documents that allow a glimpse of the role of static considerations in the construction process. Probably the most significant of these documents are the protocols of a famous consultation process in Milan at the end of the fourteenth century.

The[68] issue of this consultation between local and invited (French) master masons (plus, of course, representatives of the guilds, the city, the court etc.) concerned an ambitious change to the originally modest conception of Milan cathedral. The question at stake was: Will the foundations laid fifteen years earlier and the piers already built support a considerable increase in the building's height? Several points of the consultation are of interest in our context: First, the measurements, proportions, and form of the proposed new elevation of the cathedral were developed and designed by means of geometric grids – in one proposal a grid *ad quadratum*, in the competing one a grid *ad triangulum* –, not on the basis of static principles. Second, in the disputations with the French master masons the local masters revealed an astonishing lack of grounded understanding of statics – unfounded not only in hindsight, but also in the eyes of the French experts of the time. Third, the suggestions of the latter for mending foundations and piers reflect the rich experience of these experts but

[65] See Schlimme et al. (2014) 295ff.

[66] For these conjectures and speculations, see, for instance, Konrad Hecht (1997), W. Müller (1990) 229ff.; see also Long (2001) 213ff. and Sakarovitch (2003) 71.

[67] Some of these rules have been reconstructed from Renaissance and Baroque treatises, see Huerta (2002).

[68] For the following, see Ackerman (1991).

do not give a hint about static principles they held. The following conclusion by James Ackerman thus comes as no surprise:

[…] structure plays a secondary role in the process of creation. The overall character of the Gothic cathedral is determined on the basis of geometrical grids of lines and dots in which the specific problems of form and structure play no part. Within this embracing pattern, the actual elements of the cathedral thereafter take shape by virtue of compromise of ideal formulae and practical know-how.[69]

Almost exactly 100 years after this consultation, Leonardo da Vinci served at the court of Ludovico Sforza in Milan and, among his several other occupations, explored structural issues in particular. As the entries in his notebooks testify, he reflected static problems in a true theoretical manner. However, the conclusions he arrived at in this way remained generally unknown to his contemporaries and many generations afterwards.

These entries exhibit the double character of considerations on construction statics of the age: They are partly just geometrical rules, e.g. that an arch will not break if the chord of the outer arc does not touch the inner one (1); however, they also address questions of construction by reasoning using the methods of statics, so when Leonardo tried to determine the thrust of the stones of an arch according to their respective place in the arch he established an early version of the so-called wedge theory of arches (2)[70] – (Fig. 2.5). In view of the important role of investigations and evaluations of the static and damage to existing buildings in theoretical statics in the seventeenth and eighteenth centuries it should be noticed that cracks, break points and deformations of arches and vaults took center stage in Leonardo's examination of this topic.

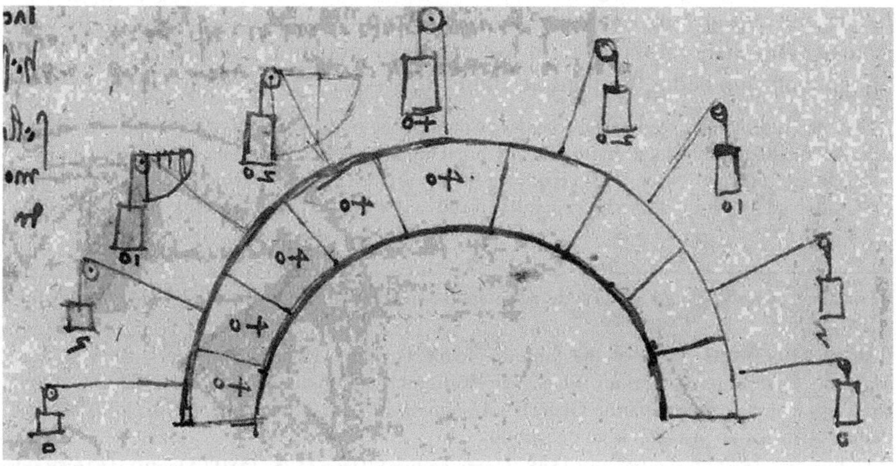

Fig. 2.5 Leonardo' Version of the Wedge Theory of Arches
Codex Madrid I – Leonardo da Vinci (1974) I, f. 142v

[69] Ibid. 246. Ackerman's article is the classic study on the Milan consultation. See also Hecht (1997) 113–170. For the geometric grids, see Velte (1951) and Lyman (1987).

[70] (1) Paris A f. 51a; (2) Madrid I f. 142v-143r. For Leonardo's studies on arches and vaults, see, for instance, Chastel (1987) or Kurrer (2002) 214 and 386ff.

Among the few documents that offer some insights into the understandings of static problems held by late Gothic master builders is one from Spain that deserves mention. In the first half of the sixteenth century, the Spanish architect Rodrigo Gil de Hontanón (1500–1577) drew up a text on how to calculate the strength of the abutment piers for brick arches, thereby presenting arithmetical as well as geometric rules. However, Hontanón's manuscript, like Leonardo's notebooks, remained unknown to most of his contemporary peers. It was incorporated in a compendium on architecture compiled in the 1680s and published only in 1868.[71]

Astonishingly, the silence regarding structural design characteristic of Gothic master builders did not end with the fading away of construction in the Gothic style. No principal change emerged in this regard, neither with the architects of the Renaissance nor even with those of the Baroque. None of the architects of the sixteenth and seventeenth centuries contributed anything of weight to the development of structural design in this period.[72] Change came from another quarter, namely that of Italian mathematical practitioners and other figures with a scholarly education who studied ancient texts on mechanical problems, particularly works by Archimedes and Hero of Alexandria, which had appeared in Latin translations produced and printed in the sixteenth century.[73]

Among the men who studied these classical texts, as well as those of Aristotle and Pappus at the turn of the sixteenth century, we find a group of physicians, mathematical practitioners, and private scholars around Federico Commandino (1506–1575) and Guidobaldo del Monte (1545–1607) that is of particular interest in our context.[74] Their investigations into the science of mechanics, comprising translations and editions of classic texts as well as theoretical treatises, created the intellectual atmosphere for the work of two men who brought the science of the strength of materials into being, Bernardino Baldi (1553–1617) and Galileo Galilei.

> Galileo is generally seen as the founder of the science of the strength of materials. His discussions of the tensile strength and the breaking strength of materials in his *Discorsi* of 1638 are regarded as the birth certificate of this science.[75] Galileo, whose commerce with Guidobaldo is well known, generally owed a lot to the achievements in mechanics

[71] For the rules presented in Gil de Hontanón's manuscript, see Kubler (1944) and Müller (1990) 235ff.

[72] Straub (1992) 107. Most of the famous architects of the Roman Baroque – such as Carlo Maderno, Gian Lorenzo Bernini, Francesco Borromini, and Pietro da Cortona – did not compile treatises on architecture. Guarino Guarini's *Architettura Civile* (drawn up c. 1670, published posthumously in Torino in 1737) is an exception.

[73] Archimedes 1543 (Tartaglia) and 1558 (Commandino), Hero 1575, 1589 (Aleotti), 1616 (Baldi) etc. Since Hero's *Mechanica*, preserved only in Arabic, was not yet published at the time, his and Archimedes' discussion of the support of a beam or architrave by pillars remained unknown to them. See Drachmann (1963) 91–146.

[74] For this group, see, for example, Drake and Drabkin (1969) 10–16 and 41–52; see also Nobis (2009).

[75] *Discorsi e dimonzationi mathematiche intorno a due nuove scienze* … (Leiden: Elzevir, 1638), First day and Second day. For Galileo's treatment of the subject, see, for example, Kurrer (2002) chap. 6.3.

accomplished by the aforesaid group. Thus, he may have known Baldi's discussion on the strength of materials.

Architecture was among the various scientific fields of interest of the polymath Bernadino Baldi,[76] abbot of the abbey of Guastalla, as is testified by his *Lexicon Vitruvianum*.[77] He developed his conceptions on the strength of materials in a rather unexpected context, namely in a commentary on Ps-Aristotle's *Mechanical Problems*, which he worked on for many years around 1600 and which was published posthumously in 1621.[78]

With regard to the theme of the present book, namely the relationship between the developments of technological and scientific literature, two questions must be posed: First, did technological literature on constructing play a part in this scientific achievement? and second, did this achievement leave traces in seventeenth-century technological literature on constructing? It seems that the answer to both questions must be no. First, there was no existing contemporary technological literature for Baldi or Galileo to draw upon. (This, of course, does not exclude both of them having knowledge of methods or rules by which master builders of the age assessed the strength of materials.[79]) Second, one reason why Baldi's and Galileo's theories left almost no traces in seventeenth-century treatises on architecture[80] might have been that questions of the strength of materials played a minor role as long as these treatises dealt with established types of buildings constructed out of well-known materials and not, for example, bridges of daring spans. Another reason was certainly the fact that these treatises mirrored the actual design practice of Italian architects of the seventeenth century.[81] And this practice followed in general.

> *[...] traditional principles of design. Geometry and proportions, plus some basic recommendations on correct and useful construction procedures were pointed out later on as well by sound architects such as Carlo Fontana, Guarino Guarini and Bernardo Antonio Vittone.*[82]

The theoretical achievements of Baldi and Galileo were first taken up, not by master builders or architects, but by men like Robert Hooke (1635–1703) and Philippe de La Hire (1640–1718) who, just like Baldi and Galileo, were not architects but experts in the mathematics and mechanics of the age. Both men delivered contributions on issues of structural design in sessions of the Royal Academies in

---

[76] For the following, see Becchi (2014b).

[77] *De verborum Vitruvianorum significatione* (Augsburg 1612).

[78] *In mechanica Aristotelis problemata excertitationes* (Mainz 1621). His conceptions on the strength of materials are developed in his commentaries on Ps-Aristotle's problems 14 and 16.

[79] In the fifth book of his treatise on fortification (see note 16 above), Buonaiuto Lorini discussed the stability of small machine models in comparison to full-scale machines, a discussion Galileo might have known about. See Büchi (2012).

[80] An exception is Cosimo Noferi's *Travagliata Architettura*, mentioned above (note 11).

[81] This holds also for François Derand's very sophisticated treatise on vaults mentioned above (note 60) that still treated structural questions geometrically since its focus was on stereometrical issues in the tradition of De l'Orme. Carlo Fontana (1638–1714), in his *Il tempio Vaticano* of 1694, provided not only the geometric profile of a cupola but also a complete set of rules for constructing cupolas. See, for instance, Schlimme et al. (2014) section 2.9.5. and Schlimme (2015) 72ff.

[82] Fusco and Villani (2003) 579.

London and Paris, respectively.[83] But this does not mean that construction statics had become a topic of a purely theoretical nature detached from the practice of architects.

> *De La Hire's theory of vaults was not just an academic exercise in classical mechanics. Its purpose was also a scientific legitimation of vault designs – that is, a justification of the* règles de l'art *by classical mechanics.* (Kurrer).[84]

Finally, in the last third of the seventeenth century we meet architects like Christopher Wren (1632–1723) or Claude Perrault (1613–1688) who had sufficient command of advanced mathematics to be able to benefit from the contemporary achievements of modern mechanics when designing structures.[85] It was these architect-scientists who initiated an era in which technological literature on structural issues and modern mechanics – particularly analytical mechanics – developed in permanent commerce.

To name just some famous contributions to this commerce from the beginning of this era: *Curvatura Laminae Elastica* (1694) by the mathematician Jacob Bernoulli (1655–1705), *La science des ingénieurs* (1729) by the engineer and architect Bernard Forest de Bélidor (1698–1761) or, later, *Memoir on statics* (1773) by the military engineer and physician Charles-Augustin Coulomb (1736–1806), and *Traité de l'art de bâtir* (1802–17) by the architect Jean-Baptiste Rondelet (1743–1829).[86]

2.8 Conclusion

According to the variety of fields of knowledge or – as Vitruvius put it – "disciplines" that make up the science of architecture, we found interrelations between the literature pertaining to some of these disciplines and scientific literature

[83] De La Hire also presented his theories in a treatise: *Traité de mecanique* (Paris 1695). See Kurrer (2002) chap. 6.4 and Becchi (2014b) 418ff. Issues in connection with statics of construction were also discussed by prominent scientists of the time, e.g. by Christiaan Huygens (1629–1695), Leibniz, Musschenbroek (1692–1761), and others; moreover, experiments concerning the strength of materials were carried out, e.g. by the architect Pierre Bullet (1639–1716), and a table with the values resulting from such experiments by the mathematician Antoine Parent (1666–1716) was published in 1718.

[84] Kurrer (2002) 215: "Die Gewölbetheorie La Hires war nicht eine rein akademische Übung der klassischen Mechanik, sondern speiste sich auch aus dem Bedürfnis nach wissenschaftlicher Legitimation des Gewölbeentwurfs – mithin der Begründung der *règles de l'art* durch die klassische Mechanik."

[85] "No quantitative application of statical theory is recorded before the time of Wren." – Mainstone (1968) 306.

[86] Jacob Bernoulli: "Curvatura Laminae Elastica …" *Acta Eruditorum* 1694 (pp. 262–276); Bernard Forest de Bélidor: *La science des ingénieurs dans la conduite des travaux de fortification et d'architecture* civile (Paris: Jombert 1729); Charles-Augustin Coulomb's *Memoir on statics* (Memoir read to the French Academy 1773); Jean-Baptiste Rondelet's *Traité théorique et pratique de l'art de bâtir* (Paris: published by the author, 1802–17).

pertaining to different sciences – and not only to natural ones, but to humanistic scholarship as well as to mathematics and mechanics.

Because of the paradigmatic significance of Vitruvius' *De architectura* for the architecture of the Renaissance and for humanistic antiquarian studies, close interchanges developed between humanistic scholarship and architectural expertise in this period, not only in relation to Vitruvius' text but also to archaeological questions concerning Roman antiquities.

As regards the relationship between erudite mathematics in the tradition of Euclid and the geometrical methods applied in the architectural design and construction processes, we confirmed that all the really remarkable methods cannot be taken as applications of erudite (Euclidean) geometry but rather as instances of a "constructive geometry" (Shelby) that developed alongside it. If there were any exchanges between these two geometries, they were unilateral: Whereas master builders like Roritzer or De l'Orme very likely took advantage of suitable constructions and theorems they could find in Euclid's *Elements*, the geometry of these practitioners left no traces in the erudite geometry taught at the universities at the time. It was not until 1600 that the mathematical potentials inherent in the stereotomic and perspectival constructions of architects and artists were recognized and theoretically developed by learned mathematicians. This was one of the rare cases in which techniques transmitted in technological writings formed the starting point of new scientific disciplines – of the non-Euclidean mathematical disciplines of projective and descriptive geometry.

Finally, as regards structural design and scientific statics we came across the strange and almost unbelievable fact that, according to our present understanding, the master builders of the Gothic as well the Renaissance age did not follow any principles of statics when designing structurally, but a number of geometrical procedures and rules instead. There was thus no existing technological literature on structural design for natural philosophers like Baldi and Galileo to refer to when addressing structural questions like the strength of materials. Accordingly, it was not until the second half of the seventeenth century that some architects began to resort routinely to early modern scientific statics when designing structures, and that a technological literature on structural design emerged. This literature then developed by permanent interchange with early modern mechanics and came to constitute a resource as well as a challenge for this science.

References

Ackerman, James S. 1991. "Ars Sine Scientia Nihil Est". Gothic Theory of Architecture at the Cathedral of Milan [1949]. In *Distance Points: Essays in Theory and Renaissance Art and Architecture*, ed. James S. Ackerman, 211–268. Cambridge, MA: The MIT Press.

Andersen, Kirsti. 2007. *The Geometry of an Art – The History of the Mathematical Theory of Perspective from Alberti to Monge*. New York: Springer.

Arnold, Thomas. 2002. *The Renaissance at War*. London: Cassell.

Barnes, Carl F. 2007. *The Portfolio of Villard de Honnecourt. A New Critical Edition and Color Facsimile*. Burlington: Ashgate.

Barzman, Karen-Edis. 2000. *The Florentine Academy and the Early State. The Discipline of Disegno*. Cambridge: Cambridge University Press.
Becchi, Antonio. 2014a. Die Gestalt der Säule. In *Wissensgeschichte der Architektur*, ed. Jürgen Renn, Wilhelm Osthues, and Hermann Schlimme, 369–396. Berlin: Edition Open Access.
———. 2014b. Architektur und Mechanik. In *Wissensgeschichte der Architektur*, ed. Jürgen Renn, Wilhelm Osthues, and Hermann Schlimme, 397–428. Berlin: Edition Open Access.
Blunt, A. 1958. *Philibert de l'Orme*. London: Zwemmer.
Böker, Johann Joseph. 2005. *Architektur der Gotik – Bestandskatalog der weltgrößten Sammlung an gotischen Baurissen der Akademie der Künste Wien*. Salzburg: Pustet.
Booz, Paul. 1956. *Der Baumeister der Gotik*. München: Deutscher Kunstverlag.
Bork, Robert. 2011. *The Geometry of Creation. Architectural Drawing and the Dynamics of Gothic Design*. Farnham: Ashgate.
Brothers, Cammy. 2017. What Drawings did in Renaissance Italy. In *The Companion to Early Modern Architecture*, ed. Alina Payne, vol. I. Hoboken: Blackwell Press.
Büchi, Tobias. 2012. "Mathematische Demonstration« und »mechanische Operation": Buonaiuto Lorini, Galileo Galilei und die Analogie von Festung und Maschine. *Scholion, Bulletin* 7: 41–57.
Bürger, Stefan. 2013. *Architectura Militaris – Festungsbautraktate des 17. Jahrhunderts von Specklin bis Sturm*. Berlin: Deutscher Kunstverlag.
Camerota, Filippo. 2004. Renaissance Descriptive Geometry. In *Picturing Machines 1400–1700*, ed. Wolfgang Lefèvre, 175–208. Cambridge, MA: The MIT Press.
Campa, Maria Rita. 2006. "Les Nouvelles Inventions pour Bien Batir et a Petits Fraiz" by Philibert de l'Orme: A New Way to Conceive Wood Roof Covering. In *Proceedings of the Second International Congress on Construction History*, ed. Construction History Society, 525–541. Newcastle: University of Newcastle upon Tyne.
Cesariano, Cesare, ed. 1969. *Di Lucio Vitruvio Pollione de architectura libri dece (1521)*. Reprint München: Fink.
Chastel, André. 1987. The Problem of Leonardo's Architecture in the Context of his Scientific Theories. In *Leonardo da Vinci – Engineer and Architect*, ed. Paolo Galluzzi, 193–206. Montreal: Montreal Museum of Fine Arts.
Coenen, Ulrich. 1990. *Die spätgotischen Werkmeisterbücher in Deutschland. Untersuchung und Edition der Lehrschriften für Entwurf und Ausführung von Sakralbauten*. München: Scaneg.
Dhombres, Jean, ed. 1992. *L'École normale de l'an III. Leçons de mathematiques. Laplace, Lagrange, Monge*. Paris: Dunod.
Drachmann, A.G. 1963. Fragments from Archimedes in Heron's Mechanics. *Centaurus* 8: 91–146.
Drake, Stillman, and I.E. Drabkin. 1969. *Mechanics in Sixteenth-Century Italy. Selections from Tartaglia, Benedetti, Guido Ubaldo & Galileo*. Madison: University of Wisconsin Press.
Duffy, Christopher. 1979/1985. *Siege War – The fortress in the Early Modern World*. 2 vols. London: Routledge.
Dürer, Albrecht. 1969. *Hierinn sind begriffen vier Bücher von menschlicher Proportion* (1528). Reprint Nördlingen: Uhl.
Fusco, Annarosa Cerutti, and Marcello Villani. 2003. Pietro da Cortona's Domes between New Experimentations and Construction Knowledge. *First International Conference on Construction History*, Madrid.
Gerbino, Anthony, and Stephen Johnston. 2009. *Compass and Rule – Architecture as Mathematical Practice in England*. New Haven: Yale University Press.
Hahnloser, Hans R., ed. 1972. *Villard de Honnecourt: Kritische Gesamtausgabe des Bauhüttenbuches ms. Fr. 19093 der Pariser Nationalbibliothek*. 2nd ed. Graz: Akademische Druck- u. Verlagsanstalt.
Haselberger, Lothar. 1999. *Essence and Appearance. Refinements in Classical Architecture*. Philadelphia: University of Pennsylvania Museum Publications.
Hecht, Konrad. 1997. *Maß und Zahl in der gotischen Baukunst*. Hildesheim: Olms.

Huerta, Santiago. 2002. The Mediaeval 'Scientia' of Structures: The Rules of Rodrigo Gil de Montanós. In *Towards a History of Construction*, ed. Antonio Becchi, Massimo Corradi, Federico Foce, and Orietta Pedemonte, 567–585. Basel: Birkhäuser.

Huppert, Ann C. 2015. *Becoming an Architect in Renaissance Italy: Art, Science and the Career of Baldassarre Peruzzi*. New Haven: Yale University Press.

Hurx, Merlijn. 2018. *Architecture as Profession. The Origins of Architectural Practice in the Low Countries in the Fifteenth Century*. Turnhout: Brepols Publishers.

Kemp, Wolfgang. 1974. Disegno. Beiträge zur Geschichte des Begriffs zwischen 1547 und 1607. *Marburger Jahrbuch für Kunstgeschichte* 19: 219–240.

Kemp, Martin. 1992. *The Science of Art: Optical Themes in Western Art from Brunellesci to Seurat*. New Haven: Yale University Press.

Kimpel, Dieter. 1983. Die Entfaltung der gotischen Baubetriebe – Ihre sozio-ökonomischen Grundlagen und ihre ästhetisch-künstlerischen Auswirkungen. In *Architektur des Mittelalters - Funktion und Gestalt*, ed. F. Möbius and E. Schubert, 246–272. Weimar: Böhlaus Nachfolger.

Köpf, Hans. 1977. *Die gotischen Planrisse der Ulmer Sammlungen*. Stuttgart: Kohlhammer.

Kruft, Hanno-Walter. 1985. *Geschichte der Architekturtheorie*. München: C.H. Beck.

Kubler, George. 1944. A Late Gothic Computation of Rib Vault Thrusts. *Gazette des Beau-Arts* 25: 135–148.

Kurrer, Karl-Eugen. 2002. *Geschichte der Baustatik*. Berlin: Ernst.

Lancaster, Lynne C. 2005. *Concrete Vaulted Construction in Imperial Rome*. Cambridge: Cambridge University Press.

Lefèvre, Wolfgang. 2004. The Emergence of Combined Orthographic Projections. In *Picturing Machines – 1400–1700*, ed. Wolfgang Lefèvre, 209–244. Cambridge, MA: The MIT Press.

———. 2010. *Picturing the World of Mining in the Renaissance – The Schwazer Bergbuch (1556)*. Berlin: MPI for the History of Science.

———. 2017. Architectural Knowledge. In *The Structures of Practical Knowledge*, ed. Matteo Valleriani, 247–269. Cham: Springer.

———. forthcoming. Architecture on Paper: The Development and Functions of Architectural Drawings in the Renaissance. In *Creating Place in Early Modern European Architecture*, ed. Elizabeth M. Merrill. Amsterdam: Amsterdam University Press.

Lepik, Andres. 1994. *Das Architekturmodell in Italien 1353–1500*. Worms: Wernersche Verlagsgesellschaft.

Long, Pamela O. 2001. *Openness, Secrecy, Authorship: Technical Arts and the Culture of Knowledge from Antiquity to the Renaissance*. Baltimore: John Hopkins University Press.

———. 2011. *Artisan/Practitioners and the Rise of the New Sciences, 1400–1600*. Corvallis: Oregon State University Press.

Lyman, Thomas W. 1987. "Opus ad triangulum" vs. "opus ad quadratum" in Mediaeval Five-aisled Churches. In *Artistes, artisans et production artistique au Moyen Âge*, Colloque international. Centre National de la Recherche Scientifique, Université de Rennes II, vol.2, 203–219. Paris: Baral Altet.

Mainstone, R.J. 1968. Structural Theory and Design before 1742. *Architectural Review* 143: 303–310.

Marten, Bettina, Ulrich Reinisch, and Michael Korey, eds. 2012. *Festungsbau: Geometrie, Technologie, Sublimierung*. Berlin: Lukas Verlag.

McGee, David. 2004. The Origins of Early Modern Machine Design. In *Picturing Machines: 1400–1700*, ed. Wolfgang Lefèvre, 53–84. Cambridge, MA: The MIT Press.

Merrill, Elizabeth M. 2017. Pocket-Size Architectural Notebooks and the Codification of Practical Knowledge. In *The Structure of Practical Knowledge*, ed. Matteo Valleriani, 21–54. Cham: Springer.

Müller, Werner. 1990. *Grundlagen gotischer Bautechnik*. München: Deutscher Kunstverlag.

Nobis, Heribert M. 2009. *Naturgesetz und Weltmaschine: die Bedeutung der peripatetischen Quaestiones mechanicae und ihrer frühneuzeitlichen Kommentare für die Grundlegung der exakten Wissenschaften*. Augsburg: Rauner.

Olschki, Leonardo. 1965. *Geschichte der neusprachlichen wissenschaftlichen Literatur*. Reprint ed., 3 vols. Vaduz: Kraus.

Pacey, Arnold. 2007. *Medieval Architectural Drawing – English Craftsmen's Methods and their Later Persistence (c.1200–1700)*. Stroud: Tempus.

Panofsky, Erwin. 1924. *"Idea" – Ein Beitrag zur Begriffsgeschichte der älteren Kunsttheorie*. Leipzig: Teubner.

———. 1997. *Perspective as Symbolic Form (Die Perspektive als symbolische Form, 1927)*. New York: Zone Books.

Potié, Philippe. 1996. *Philibert de l'Orme. Figures de la pensée constructive*. Marseille: Parenthèses.

Rabasa-Díaz, Enrique. 2000. *Forma y Construccíon en piedra. De la cantería medievale a la estereotomía del sieglo XIX*. Madrid: Ediciones Akal.

Rabasa-Díaz, Enrique, and José Calvo-López. 2009. Gothic and Renaissance Design Strategies in Stonecutting. In *Creating Shapes in Civil and Naval Architecture. A Cross-Disciplinary Comparison*, ed. Horst Nowacki and Wolfgang Lefèvre, 167–192. Leiden: Brill.

Recht, Roland. 1989. *Les Bâtisseurs des cathedrales gothiques – Catalogue d'exposition*. Strasbourg: Musées de la ville de Strasbourg.

Roth, Leland M. 1993. *Understanding Architecture: Its Elements, History and Meaning*. Boulder: Westview Press.

Sakarovitch, Joel. 1998. *Épure d'architecture. De la Coupe des pierres à la géométrie descriptive – XVIe–XIXe siècles*. Basel: Birkhäuser.

———. 2003. Stereotomy, a Multifacetd Technique. In *Proceedings of the First International Congress of Construction History*, ed. Santiago Huerta, 69–79. Madrid: Instituto Juan de Herrera.

Schlimme, Hermann. 2006. Between Architecture, Science and Technology: The Accademia della Vacchia in Florence, 1661–1662. In *Practice and Science in Early Modern Italian Building*, ed. Hermann Schlimme, 61–96. Milan: Electa.

———. 2015. Kontexte der Lehrbuchproduktion im Italien der frühen Neuzeit. In *Der Lehrbuchdiskurs über das Bauen*, ed. Uta Hassler, 66–77. Zurich: vdf Hochschulverlag.

Schlimme, Hermann, Dagmar Holste, and Jens Niebaum. 2014. Bauwissen im Italien der frühen Neuzeit. In *Wissensgeschichte der Architektur*, ed. Jürgen Renn, Wilhelm Osthues, and Hermann Schlimme. Berlin: Edition Open Access.

Schoeller, Wolfgang. 1989. Ritzzeichnungen – Ein Beitrag zur Geschichte der Architekturzeichnung im Mittelalter. *Architectura* XIX: 36–61.

Schuler, Stefan. 1999. *Die Rezeption von "De architectura" von der Antike bis in die frühe Neuzeit*. Cologne/Weimar/Vienna: Böhlau.

Scotti, Auror. 1983. Il collegio degli architetti, ingegneri e agrimensori tra il XVI e il XVIII secolo. In *Construire in Lombardia – Aspetti e problemi di storia edilizia*, ed. Aldo Castellano and Ornella Selvafolta, 92–108. Milan: Electa.

Semper, Gottfried. 1863. *Keramik, Tektonik, Stereotomie, Metallotechnik*. Munich: Friedrich Bruckmann.

Shelby, Lon R. 1972. The Geometrical Knowledge of Mediaeval Master Masons. *Speculum* IV: 395–421.

Shelby, Lon R., and Robert Mark. 1979. Late Gothic Structural Design in the 'Instructions' of Lorenz Lechler. *Architectura* 9 (2): 113–131.

Straub, Hans. 1992. *Geschichte der Bauingenieurkunst*. Basel: Birkhäuser.

Thoenes, Christof. 1993. Vitruv, Alberti, Sangallo. Zur Theorie der Architekturzeichnung in der Renaissance. In *Hülle und Fülle – Festschrift für Tilmann Buddensieg*, ed. Andreas Beyer, 379–391. Weimar: VDG.

Valeriani, Simona. 2006. *Kirchendächer in Rom*. Petersberg: Imhof.

Velte, Maria. 1951. *Die Anwendung der Quadratur und Triangulatur bei der Grund- und Aufrissgestaltung der gotischen Kirchen*. Basel: Birkhäuser.

Chapter 3
Chemistry

3.1 Literature on Chemistry before 1600

Looking at the period before the seventeenth century, it is not at all obvious what can and should be regarded as chemical/alchemical literature.[1] According to our present understanding of chemistry – and not that of the historical actors – a wide variety of manuscripts and printed texts of this age can be classified under the unifying label "chemical literature." This is more or less clear as regards literature on technical procedures such as artists' manuals (*Kunstbüchlein*) on the preparation of colors and tints, on gilding, producing stained glass, enameling, etc., manuscripts or booklets on brewing, viniculture, cheesemaking etc. It also applies to texts on pottery and glazing, mixing gunpowder, and so on, as well as the plethora of *books of secrets* printed in the long sixteenth century.[2] These texts were written or composed and disseminated by various practitioners, including publishers, who had little in common: painters, sculptors, goldsmiths, glass makers, brewers, wine-growers, potters, artillerists, etc.

As most of these practical fields had been known and cultivated in classical as well as pre-classical antiquity and in the Arabic Middle Ages, the mediaeval authors of the West could often build on manuscripts transmitted from these civilizations, e.g. texts on arts such as the *Codex Lucca* (ninth century), Heraclius' (?-?) *De coloribus et artibus Romanorum* (tenth century) or the *Mappae Clavicula* (eleventh century).[3] Cennino Cennini's (?1370-1440?)

The question marks before, after, or instead of annual figures mean, that these figures are uncertain or unknown.

---

[1] As is well known, our distinction between chemistry and alchemy cannot be applied to texts written before the eighteenth century.

[2] For 16th- and seventeenth-century books of secrets, see, for instance, Eamon (1994).

[3] For these medieval manuscripts and their sources, see the overview on ancient as well as medieval texts on colors and painting by Christiana Herrington in the foreword to her translation of Cennini's *Libro* – Cennini (1899) xvff.: they range from the *Leyden Papyrus* (Egypt fourth century B.C.) to Theophrastus, Vitruvius and Pliny the Elder (on mural art), to the *Codex Lucca* (Biblioteca

W. Lefèvre, *Minerva meets Vulcan: Scientific and Technological Literature – 1450-1750*, Archimedes 60, https://doi.org/10.1007/978-3-030-73085-7_3

Libro dell'arte o trattato della pittura (end of the fourteenth century), for instance, probably the most influential manual on painting of the late Middle Ages, is one among several manuscripts of the time that drew on and transformed this body of transmitted knowledge.[4] Similarly, agrarian literature such as the sixth book of Vincent of Beauvais' encyclopedic *Speculum maius* (1256), Walter of Henley's *Le Dite de Hosebondrie* (1280) or Pietro de' Crescenzi's *Ruralia commoda* (1305) could rely on ancient texts like Marcus Porcius Cato's *De agri cultura* (second century BC) and Columella's *De re rustica* (first century AD), since these texts had been transmitted to the medieval West.[5] The same holds for literature about glassmaking and glass working or pottery.

Even the literature on the two technical fields which shaped the development of chemistry in the early modern period – namely pharmaceutical distillation and metallurgy in the context of mining, smelting and assaying – did not constitute an integrated body of chemical technological literature. Rather, literature on these fields existed and developed almost unconnectedly side by side until the end of the sixteenth century. We will return below to the literature of these two important practical domains.

What has been observed with regard to the diversity of crafts or professions of the authors of texts on technical procedures also applies, though to a lesser extent, for the authors of texts of a more theoretical nature, including physicians, pharmacists, smelters and master minters (*Münzmeister*), practical mathematicians (*Rechenmeister*), clergymen, schoolmasters, etc. On the other hand, and in contrast to all of its diversity we find the all-embracing field of *alchemy*: What the historical actors designated as *alchemy* united practices of a true chemical nature (as we understand them, and regardless of their efficiency), such as distillation or assaying, with theories that have (from our viewpoint) nothing to do with chemistry at all, such as general natural philosophy, astrology/astronomy or psychology.[6]

Capitolare Lucca, codex 490), a compilation of translations from Byzantine sources, the *Mappae Clavicula*, which draws on the *Codex Lucca* as well as on the *Leyden Papyrus*, Heraclius' *De coloribus*, a book with texts of various medieval authors and sources, and up to Theophilus Presbyther (twelfth century), Cennini etc.; for more recent accounts, see Long (2001) 82ff. and Clarke (2001).

[4] For the impact of Cennini's *Libro*, see M. Merrifield (1999). Other such late-medieval manuscripts on painting and colors include the *Liber de coloribus faciendis*, BNF Paris Ms. lat. 6741 (13th/fourteenth century) and the *Liber de coloribus illuminatorum*, British Library London MS Sloan 1754 (fourteenth century). Booklets on this subject which draw freely on these manuscripts were composed and published well into the second half of sixteenth century, e.g. the *Liber illuministarum* held in the monastery of Tegernsee (c. 1500), *Allerhand Farben* (1533), or the *Illuminir-Buch künstlich alle Farben zu machen* (1550, 1552, etc.). For a long-term overview of texts on paints and other painters' materials, see Laurie (1910); see also the bibliography compiled by Klaus-Peter Schäffel: http://www.schäffel.ch/download/pdf/bibliographie.pdf

[5] Medieval mss. of Columella's manual can be traced back to the ninth century, those of Cato's work to the twelfth century. For the transmission of ancient agricultural treatises, see Rex (2001). Agrarian literature will not be a topic in the following.

[6] As is well known, many of the medieval and early modern alchemical manuscripts and books do not give plain descriptions or explanations of the chemical operations they deal with. Instead, they use allegorical or pictorial intimations intelligible only to adepts. (Prominent instances are *Splendor Solis* (c. 1530), *Rosarium philosophorum* (1550) and Michael Maier's *Atalanta fugiens* (1617). Attempts at publicizing alchemical knowledge and freeing it from this guise, as well as from widespread charlatanry, such as Isabella Cortese's *I secreti della signora Isabella Cortese*

This state of affairs of the chemical literature before 1600 reflects nothing else than the state of affairs of chemistry itself in this period. There existed no united or connected field of chemical practices or theories, no standardized training in chemistry, neither by schooling nor by apprenticeship, no established and well demarcated community and profession of chemists, no standard canon of chemical literature, let alone chemical periodicals. And yet, as we will see, it was not least the literature of the sixteenth and early seventeenth centuries that contributed in a crucial way to the integration of this various fields of practices into a more and more coherent chemical technology.

3.2 Literature on Distilling

Manuscripts and books on the art of distilling represent a major domain of chemical practices in the period before 1600. The production of medicines constituted the center of this domain, which could thus be called pharmaceutical chemistry. Initially, the medicines obtained by this practice were almost exclusively distillates of vegetable and some animal materials. Since about 1500, however, more and more distillates of mineral materials emerged due to radical changes in medical views (inspired by Paracelsus and his followers) as well as to developments in ore mining and smelting. Thus, the early modern art of distilling must be seen in its manifold relations to two clusters of practical domains – growing/gardening/botany, pharmacy/medicine on the one hand and, on the other, mining/metallurgy/mineralogy; and the literature on the art of distillation should be seen in this light.

The art of distilling is an age-old art, and classical and pre-classical ancient knowledge of it as well as that of the Arabic Middle Ages had been transmitted to the medieval West. Evidence of medieval knowledge of this art is provided, for instance, by a manuscript from c. 1350 authored by the Franciscan Johannes de Rupescissa (Jean de Roquetaillade), which focuses on distilling alcohol (*aqua vitae*).[7] As regards the distillation of medicines, the transmission of the *Materia medica* of the Greek physician Pedanius Dioscorides (first century AD) deserves special attention.[8] This work is equally interesting both as a botanical and a pharmaceutical manual. This dual character is typically found in many other sixteenth-century herbals (*Kräuterbücher*) and books on distilling.

(1561), remained an exception at the time. These "texts" constitute a body of technical or theoretical chemical literature which we will not explore in the following. Rather, it is works like Libavius' *Alchemia* (1597) and Crollius' *Basilica Chymica* (1609) where we find descriptions and discussions of alchemical practices as well as theories that have a bearing on the co-evolution of technical and theoretical chemical texts.

[7] In the sixteenth century the text was published in print under the title *De consideratione quintae essentiae* (Basel 1561). Late-medieval works on distillation include Gabriel von Lebenstein's manuscript *Von den gebrannten Wässern* (HAB Code 54 Aug, c. 1400) and Michael Puff von Schrick's *Büchlein von den ausgebrannten Wässern* (Augsburg 1477).

[8] Illustrated manuscripts of Dioscorides' *Materia medica* were copied in Greek, Arabic, and Latin throughout the medieval period. The first printed edition (in Latin) was published in 1478.

The *Kleines Destillierbuch* by the physician Hieronymus Brunschwig (c. 1450 – c. 1512) is a case in point.[9] A work on procedures and understandings of distilling with a considerable impact on the development of this literary genre, it is rated at the same time among the few important botanical works that preceded the famous sixteenth-century series of botanical/pharmaceutical herbals composed by Otto Brunfels (1488–1534), Hieronymus Bock (1498–1554), Leonard Fuchs (1501–1566), and Rembert Dodoens (1517–1585).[10] For in the second book of this work, Brunschwig provided hundreds of short monographs of plants from which distillates were obtained, and added to each a woodcut image of the plant.[11] While brief notes on the medical-pharmaceutical whereabouts of each of these distillates are listed in the third book, the first book is dedicated to detailed descriptions of the distilling equipment – ovens, vessels, and other tools – and to its handling, that is, to the distilling technology. As in the case of the plants, the description of the equipment is supported by images (Fig. 3.1).

Brunschwig did not pretend to have invented or improved on the equipment he presented. Since his foreword invokes physicians, skilled master craftsmen ("gute meyster") and other experts as informants, his descriptions can probably be taken as a portrayal of the equipment distillers used at the time. Naturally, this does not exclude him drawing on previous manuscripts on this subject such as the *Liber florum Geberti*, composed by an unknown author only a quarter of a century earlier which, in turn, drew on old Byzantine and Arabic sources.[12] Brunschwig did not content himself with descriptions of the various items of distillation equipment but also gave information about the fabrication of some of them, e.g. of certain ovens. Just as important as these descriptions of the appliances of distillers are those of the various techniques and procedures they employed.

Brunschwig[13] distinguished between procedures "without" and "with expenses" ("ohne" and "mit Kosten"). The former comprise procedures of filtering and several methods of obtaining liquids from the material at hand just by warming it up. The latter comprise distilling procedures in the proper meaning of the term, that is, procedures of separating components of a mixture by selective boiling and condensation. There are detailed descriptions

[9] *Liber de arte distillandi de Simplicibus. Das buch der rechten kunst zü distilieren die einzigen ding* (Strasburg 1500); it was translated into Dutch (1517), English (1527), and Czech (1559); an enlarged edition appeared in 1512 and was reissued several times.

[10] Otto Brunfels: *Herbarum vivae eicones*. 3 vols., Strasburg: Hans Schott, 1530–1536; Hieronymus Bock: *Das Kreütter Buch*. Strasburg: Wendel Rihel, 1539; Leonard Fuchs: *New Kreüterbuch*. Basel: Michael Isingrin, 1543; and Rembert Dodoens: *Cruydeboeck*. Antwerp: Van der Loe, 1554. Two influential herbals before Brunschwig should be mentioned: Vitus Auslasser's herbal (1479), BSB clm 5905, a manuscript containing 198 plant monographs with watercolor images by Erhard Reewiijk; and Johann Wonnecke von Kaub's *Gart der Gesundheit = Herbarius zu Teutsch* (Mainz 1485) with woodcut images. For illustrated herbals, see Arber (1912) and Anderson (1977).

[11] For the presumed origin of these images, see Forbes (1948) 114 f.

[12] *Liber florum Geberti* (c. 1475), BSB Munich Clm 25,110. For the dating of the manuscript, see Ganzenmüller (1941). For a reconstruction of some of the ovens and vessels described in this manuscript, see https://www.uni-frankfurt.de/59035867/Chemie

[13] For the following, see https://de.wikipedia.org/wiki/Kleines_Destillierbuch

Fig. 3.1 Distillation Oven (1500)
Hieronymus Brunschwig: *Liber de arte distillandi* I, chap. vi, f. 4v

of how the material at hand is heated within a vessel (the *curcubit*) on which a collecting vessel (the *alembic*) is set, in which the emitted gaseous substances condense, drip out and run through a pipe into another collecting vessel. The five procedures described by Brunschwig differ with regard to the regulation of the degree of heat that is applied – in ascending order: *distillatio per balneum marie* (the *curcubit* being placed in a hot water bath), *per ventrem equi* (in a hot water bath with horse dung), *per cinerem* (in hot cinders), *per arenam* (in hot sand), and finally *per ignem* (on a fire). Continuous (fractional) distilling is not among the procedures described. The methods by which the materials are prepared for distillation are only briefly addressed. Whether or not solvents (water, alcohol or vinegar) had to be applied is usually not specified.

Among the books on distilling that followed Brunschwig's in the long sixteenth century we find many compilations that drew on his book, especially on the enlarged edition of 1512, as well as on texts by other authors. We should take note of these compilations since they certainly contributed to disseminating pharmaceutical knowledge.[14] Of that century's more serious books on distilling, such as those by Philipp Ulstadt (?-?), Adam Lonicer (1528–1586), Pietro Andrea Mattioli (1501–1577), Geronimo Rossi da Ravenna (1539–1607), Giambattista Della Porta (1535–1615),

[14] *Liber de arte Distillandi de Compositis. Das Buch der waren Kunst zu distillieren die Composita und Simplicia, und das Buch Thesaurus Pauperum.* Strasburg: Johann Grüninger, 1512. The following can be regarded as derivates of this edition: *Das neue Destillierbuch* (Straßburg 1519), *Das Buch zu Destillieren* (Straßburg 1519 u. 1531), *Das Destillierbuch der Rechten Kunst* (Frankfurt 1553), *Hausarzneybüchlein* (Frankfurt 1594), *Thesaurus Pauperum* (Frankfurt 1598), and the Appendix of Peter Uffenbach's *Kräuterbuch* (Frankfurt 1610), which was a reissue of Adam Lonicers' herbal. A notorious compilation is Walther Ryff's *Das new groß Destillier-Buch* (1545).

and Jean Liébault (1535–1596),[15] one merits particular mention, namely Konrad Gessner's (1516–1565) *De remediis secretis* (1552), which is regarded as the most comprehensive sixteenth-century book on distilling, written by a physician renowned above all for his zoological works.[16] It reminds us to bear in mind that not only vegetable and mineral materials were subjected to distillation, but animal materials as well.

3.3 Literature on Assaying and Smelting

Metallurgy, assaying and smelting was the other major domain of chemical practices in the period before 1600. It developed and flourished during the early heyday of metal mining in the West, in an age of advanced artillery warfare and a growing economic and political importance of metallic currency and, along with this, the control of coinage alloys. Accordingly, we find the authors of literature on these practices among mint masters (*Münzmeister*) like Lazarus Ercker (1528–1594) and Modestin Fachs (?-1575), and mining and smelting experts like Vannoccio Biringuccio (1480–1537), as well as among physicians like Georgius Agricola (1494–1555) who were driven by pharmaceutical and medical interests in minerals, particularly metals. Although the art of smelting and assaying is as old as the art of distilling, it seems that sixteenth-century authors on metallurgy hardly had any recourse to medieval manuscripts that transmitted practical knowledge on assaying and smelting from Classical Antiquity or other civilizations.

There are a few medieval manuscripts that deal with metallurgy in the context of metalworking: Two manuscripts from the early Middle Ages – the *Codex Lucca* and the *Mappae clavicula* (both from the ninth century) – provide recipes on metalworking that stem from ancient and Arabic sources. Theophilus Presbyter, in the foreword of his *De diversis artibus* (twelfth century), refers to unspecified Arabic sources on metalworking. This manuscript certainly had no significance for writers of the sixteenth century, since it was only discovered in the second half of the eighteenth century. The *Summa perfectionis magisterii* (Pseydo-Geber) from the thirteenth century, a Latin manuscript with Arabic roots, describes a cupellation procedure, that is, an operation in the context of ore smelting. A section on reinforcing iron in a Nuremberg *Hausbuch* (dated 1389) contains no reference to Greek, Latin, Arabic or other sources; instead it begins with the words: "Nu spricht meister Alkaym" ("Now speaks Master Alchemy"). It is an open question whether any of these manuscripts was known to sixteenth-century authors on metallurgical topics.[17] In contrast,

---

[15] Philipp Ulstadt: *Coelum philosophorum* (1526), Appendix on distilling in Adam Lonicer's *Kräuterbuch* (1557), the *Appendix* of Pietro Andrea Mattioli's *Commentary on Dioscorides' materia medica* (1544, 1554), Geronimo Rossi da Ravenna: *De Destillatione Liber* (1582), Giambattista Della Porta: *De Distillationibus, Libri IX* (1608), and Jean Liébault: *Quatre livres des secrets de médecine et de la philosophie chymique* (1573). For more details, see Forbes (1948).

[16] Gessner, Konrad: *Thesaurus Evonymi Philiatri de remediis secretis*. Zurich: Gessner, 1552, reprinted several times. See Forbes (1948) 121ff. and Fischer (1966) 78ff.

[17] *Codex Lucca* and *Mappae clavicula*: Lucca, Biblioteca Capitolare Feliniana, Codex 490; Theophilus Presbyter's *De diversis artibus*: Vienna NB Cod. 2527 and HAB Cod. Guelf. Gud. Lat. 69 2°; *Hausbuch:* German Nationalmuseum Nürnberg, Cod. MS 3227a, fol. 11r-12r.

> ancient and Arabic *mineralogical* writings on metals, e.g. those of Aristotle or Avicenna, were well known to the medieval West and even further developed by authors like Albertus Magnus, on whose works early modern metallurgists as well as mineralogists were able to draw.[18]

Against this background, the degree of maturity of sixteenth-century literature on metallurgical techniques is all the more remarkable. The very first Western treatise on metallurgy, Biringuccio's *De la Pirotechnia,* is already an amazing book.[19] Its author, Vannoccio Biringuccio, was chief supervisor of the mines and head of the mint of the Republic of Siena. The treatise is extraordinarily comprehensive.

> It starts by dealing with the deposits and the mining of ores (book I); gives descriptions of a wide range of minerals – not only ores and metals and some half metals but also sulfur, arsenic, alum etc. (book II); goes into the details of the preparation of ores for smelting (book III) and discusses various smelting procedures along with descriptions of how to construct and operate the ovens employed (book VII); and describes, moreover, methods of separating gold and silver (book IV) and of alloying copper, lead, and tin (book V). It then proceeds to discuss various arts and crafts of metalworking such as metal casting, particularly gun barrel casting (book VI), and, as could be expected, mintage (book IX); and it ends with pyrotechnics and related matters (book X).[20]

It is not just the extensive scope of Biringuccio's portrait of early modern metallurgical practices that is amazing. By discussing the knowledge pertaining to all of the various practices of miners, smelters, assayers, and the various kinds of metalworkers, his treatise revealed the layout and structure of metallurgy as a connected body of practical knowledge that interconnects heterogeneous kinds of practical knowledge. In doing so, Biringuccio's treatise accomplished no less than to lay the foundation of metallurgy as a connected field of practical knowledge, and established itself as this field's first manual.[21]

---

[18] Aristotle: *Meteorology* III.6, Avicenna: *De congelatione et conglutinatione lapidum*, Albertus Magnus: *De mineralibus et rebus metallicis*.

[19] *De la pirotechnia: libri x … composti per il s. Vanoccio Biringuccio Sennese*, published (posthumously) Venice: Venturino Rufinelli, 1540. For Biringuccio and his treatise, see introduction to the English edition of the *Pirotechnia* (Biringuccio 1959) by Cyril S. Smith and Martha T. Gnudi; see also Olschki (1965) II 269ff.; Partington (1961–70) II 32ff.; Klein (1994) 109 f.; P. Long (2001) 178ff, and Bernardoni (2011).

[20] Books 7 through 11 of Georgius Agricola's *De re metallica* of 1756 provide similarly comprehensive descriptions and explanations of techniques and tools of metallurgists in connection with mining and smelting. Moreover, these books are lavishly illustrated. We will return to Agricola below in the chapter on mining science (*Bergbauwissenschaft*).

[21] It must be added, though, that as early as 1524, a *Probierbüchlein* appeared in Germany that seems to be a compilation of recipes of unknown provenance. Fifteen more editions of it appeared in the course of the sixteenth century: e.g. *Probir buchleyn: auff Golt, Silber* … s.l., 1527. For a modern English translation and edition of this booklet, see Sisco and Smith (1949). Further compilations of texts on metallurgical issues were published in the sixteenth century, e.g. Anonymous: *Bergwerck und Probirbüchlin* (Frankfurt: Christian Egenolph, 1533) that comprises a shortened version of Rülein von Calw's *Bergbüchleyn* of 1505 with texts from a *Probierbüchlein* by an anonymous author.

Metallurgy, as outlined in this first manual, exhibits characteristic features of practical knowledge pertaining to complex fields of manufacturing and production such as architecture/construction or mining. The practical knowledge of such fields consists of the competences of the various actors cooperating in the field. The structure of metallurgy as a technological science is thus not a mental structure integrated by a theory; rather, it is the fabric of fragmented pieces of knowledge connected by the cooperation of different players with different competences and different social profiles – craftsmen from several branches as well as persons with some formal education or those able to benefit from manuals like Biringuccio's or other literature. Among them we find practitioners able to coordinate and control a considerable number of these diverse competences based on broad experience and expertise in several (not all) of the practices that constitute this field. As to metallurgy, it was particularly the assayers who played such a central role in the mining and smelting business.[22]

> Decisions on whether or not it was economically worth mining an ore deposit depended on the assaying of the ore at hand. This assay's results determined the choice and composition of additional substances needed for the smelting of the ore. Moreover, the assayer had to control the composition of the extracted alloys several times during the smelting process. His competence as regards the procedures of smelting was based on the fact that assaying comprised all smelting techniques on a smaller scale. Finally, it goes without saying that assaying in connection with minting was a particularly responsible task that gave the assayer an important status in the social fabric of metallurgy as well as beyond.

An assayer not only had to have specific metallurgical knowledge and expertise at his command. He also had to be conversant with several crafts pertaining to the assaying and smelting processes, e.g. with the making and, if needed, repair of the appliances employed in the business. In the introduction to his *Aula subterranea* (1573), the mint master Lazarus Ercker enumerated "what an assayer should know"[23]:

> *First in importance is a knowledge of metallic ores and minerals, namely how to distinguish one from another by their appearance and color. [...]*
>
> *[...] second [...] knowledge of fire. The assayer should know how to regulate it so that no metal is exposed to it more than is necessary [...]*
>
> *He himself should be able to make and prepare skillfully all his own furnaces and tools [...] or he should at least be able to give the right directions for making them [...]*[24]

[22] For the following, see Klein (1994) 106 f.

[23] Sisco and Smith (1951) 10 f.

[24] "*To prepare for emergencies, every assayer ought to be able to make his own assay furnaces, scorifiers, crucibles, muffles, and such utensils as are needed daily in assaying, since a master who knows how to do it cannot be found everywhere. Even if the assayer spends much effort in instructing the potters [...] it happens only too often that they produce something that is neither of the right quality nor of the right shape [...]*" Ibid. 24.

> *He should insist on having good, sensitive balances and correct weights and should make and adjust them himself, if necessary.*[25]

The issue of assayer's balances and weights deserves particular attention. If the capacity of adjusting, calibrating, and, if needed, repairing the assayer balances and weights belonged to the expected qualifications (that is, the informal qualification profile) of mint masters, we must rank them among mathematical practitioners such as surveyors or navigators who had to meet the same (informal) profile requirement with regard to the mathematical instruments they used. Accordingly, Ercker concludes his listing of an able assayer's competences by adding[26]:

> *In addition to mastering the items and points so far enumerated, the assayer should be well versed and experienced in arithmetic or calculation; [...] indeed, it is the very sign of a master.*

It is therefore not surprising that we encounter a computing or accounting master (*Rechenmeister*) among the authors of sixteenth-century *Probierbüchlein*, namely Zacharias Lochner (c. 1557–1608).[27] Regarding the whole spectrum of qualifications of mint masters – mineralogical knowledge, expertise in all procedures of smelting and assaying, expertise in the fabrication, operating, and, if necessary, repair of the equipment of smelting and assaying, and expertise in the mathematics required for weighing and quantitative analyses – mint masters resemble the engineers of the age who combined practical knowledge with an educational background. This *persona* of the engineer played a pivotal role for the trade or commerce between technological and scientific literature in the early modern period.

3.4 The Emergence of Chemical Technology

Although the technical literature of the sixteenth century presents pharmaceutical and metallurgical operations as two separate fields of chemical practices, this separation broke down step by step in that same century. At first, points of contacts between the fields of pharmaceutical and metallurgical practices developed as physicians began to include more and more preparations from mineral substances in the arsenal of applicable medicines. The Swiss physician Paracelsus, born Theophrastus von Hohenheim (c. 1493–1541), is seen as the most instrumental figure in this

[25] "*An assayer not only possesses very neatly made and true assay balances but should also know how to repair them and what to do in order to correct them [...] If they want to be sure of their assays and able to rely on them, they should really be skillful enough to make their own balances, assay weights, and assay tools, as well as to adjust them.*" Ibid. 82.

[26] Ibid. 11.

[27] Zacharias Lochner: *Probier-Büchlein*. Augsburg: 1565. Beside the books of Ercker and Lochner, two other sixteenth-century *Probierbücher* are worth mentioning: Samuel Zimmermann: *Probier büch: Auff alle Metall*. Augsburg: M. Manger, 1573; Modestin Fachs: *Probier Büchlein*. Leipzig: Barwald u. Grosse, 1595.

development. The label *iatrochemist* was attached to him and his followers exactly because they extended the realm of medicines beyond distillates of vegetables to include mineral preparations. To implement such preparations as medicines these physicians and pharmacists had to discover, process, and test mineral substances, in particular metals such as mercury und antimony, that is, they had to apply procedures and practices of the metallurgical chemical field.

In contrast to other alchemical authors of the age, iatrochemists usually openly disseminated their knowledge about chemical operations for achieving various medicines and preparing them from metals and other mineral substances. Physicians and pharmacists could use the writings of Basil Valentine (Basilius Valentinus (?-?)) or the physician Oswald Crollius (1560-1609), for instance, as manuals about this part of their business.[28] In the course of the seventeenth century, iatrochemists were appointed as professors of medicine and iatrochemical chemistry was taught at universities. Chairs for iatrochemistry were even established at two German universities: in 1609, Johann Hartmann (1568-1631) was appointed as professor for *Chymiatria* at the University of Marburg and, in 1641, Werner Rolfinck (1599-1673) as professor for iatrochemistry at the University of Jena.[29] The appointment of the Dutch iatrochemist Franciscus Sylvius, born Franz de le Boë (1614-1672), by the famous medical faculty at the university of Leyden actually led to the establishment of a chemical laboratory.[30]

Two points must be kept in mind for a proper understanding of this transformation process of pharmaceutical and metallurgical chemistry into parts of one comprehensive chemical technology. *First*, the growing number of chemical medicaments did not displace those produced by distilling vegetable materials. As these newcomers supplemented rather than displaced the latter, metallurgical operations came to supplement, not displace, distilling in pharmaceutical chemistry. In contrast to iatrochemical medical doctrines, which partly ousted the old Galenic ones, the new chemical medicines did not abolish the traditional ones. And just as traditional medicines proved to be of more viability than traditional medical doctrines, in turn the new chemical medicaments survived the decline of the medical doctrines of Paracelsus and his followers in the second half of the seventeenth century. *Second*, pharmaceutical and metallurgical chemistry did not simply merge in this process. Not all of the procedures of assaying and smelting became practices applied by physicians and pharmacists; and conversely, most of their distilling procedures did not become techniques applied by assayers and smelters. The two domains of chemical practice overlapped in a particular segment of metallurgical operations, namely

[28] Basil Valentine, an author whose identity is still uncertain, published many texts on iatrochemical issues around 1600, among them probably the most read *Zwölf Schlüssel* (included in in his *Ein kurtz Summarischer Tractat,* Eisleben 1599) and *Triumphwagen Antimonii* (Leipzig 1604). Oswald Crollius' *Basilica Chymica* (Frankfurt 1609), written in Latin, translated into German and other vernaculars and reissued several times, was a handy compendium of the medical as well as chemical doctrines of Paracelsus and his followers. See Debus (1977) I 94 f. *ad* Basil Valentine and 117ff. *ad* Crollius.

[29] For Hartmann, see Moran (1991) 13ff.; for Rolfinck, see Jaumann (2004) 565.

[30] See Underwood (1972).

in processing metals (ores, alloys, metallic oxides and metals proper) as well as other minerals by means of acids.

The emergence of a comprehensive chemical technology did not remain a gradual and mute process, that is, a process resulting solely from the actors' deeds. Rather, the comprehensiveness and connectedness of chemical technologies practiced at the turn of the sixteenth century suddenly became visible and explicit when Andreas Libavius (c. 1555–1616) and Jean Beguin (1550–1620) published comprehensive chemical treatises that presented and reflected this connected field of chemical practices alongside the understandings accompanying them.[31] Both books[32] are rightly regarded as the very first manuals, and even textbooks, of chemical technology. We could say that chemical technology, that is, a comprehensive science of chemical procedures and techniques, first came into being with these manuals. Both treatises, besides dealing with the whole range of dissolution and distillation techniques, also treat several metallurgical procedures such as calcination and sublimation. In the first part of his manual, the *Encheria* or instruction on treatment, Libavius also provided descriptions of the various items of equipment used by chemists of the time. These descriptions remained unsurpassed until the publication in the mid-seventeenth century of Johann Rudolf Glauber's (1604–1670) *Furni novi philosophici* (1646–49).[33]

The two authors did not confine themselves to giving descriptions and explanations of the various ways and means by which substances were processed and transformed by pharmacists, metallurgists, and other chemical practitioners such as glassmakers. They tried to delineate the structure of this vast field of chemical practices in a comprehensive and systematic manner. In order to render this structure easily understandable, both authors used not only descriptions and deliberations but also instructive diagrams. (Fig. 3.2).

> These diagrams illustrate the relationships of the various chemical operations independent of the particular fields of chemical practice in which they are applied. The diagrams demonstrate these relationships by way of classification, that is by ordering chemical operations into categories regardless of the special practical fields in which they are applied. For example, Libavius' diagrams show that *Sublimatio* (sublimation) and *Distillatio* (distillation) are classified as kinds of *Prolectatio* (elicitation) and the latter together with *Expressio*

---

[31] Andreas Libavius, studied philosophy, history, and medicine, and became city physician and headmaster of the high school (Gymnasium) in Rothenburg ob der Tauber. Jean Beguin, who had been apprenticed to a pharmacist and had also studied medicine, ran a laboratory for medicines in Paris; he visited mining districts in Italy, Germany, and Hungary. See Moran (2007) and Debus (1977) I 169ff.

[32] Andreas Libavius: *Alchemia*, Frankfurt 1597; two volumes with teaching letters of 1595 preceded it and two supplementary volumes with commentaries were published, in 1606 and 1611 respectively. Jean Beguin: *Les élémens de chymie*, Paris 1620. A shorter version in Latin (*Tyrocinium Chymicum*) was already published by Beguin in 1610. See Partington (1961–1970–70) II 244ff., III 2ff. and Multhauf (1961).

[33] In the commentary published in 1606, Libavius included about two hundred figures picturing the various items of equipment used by chemists; these figures are reproduced as an appendix to the modern German edition of the *Alchemia* (Weinheim 1964). Johann Rudolph Glauber: *Furni novi philosophici.* Amsterdam: J. Fabel, 1446–49.

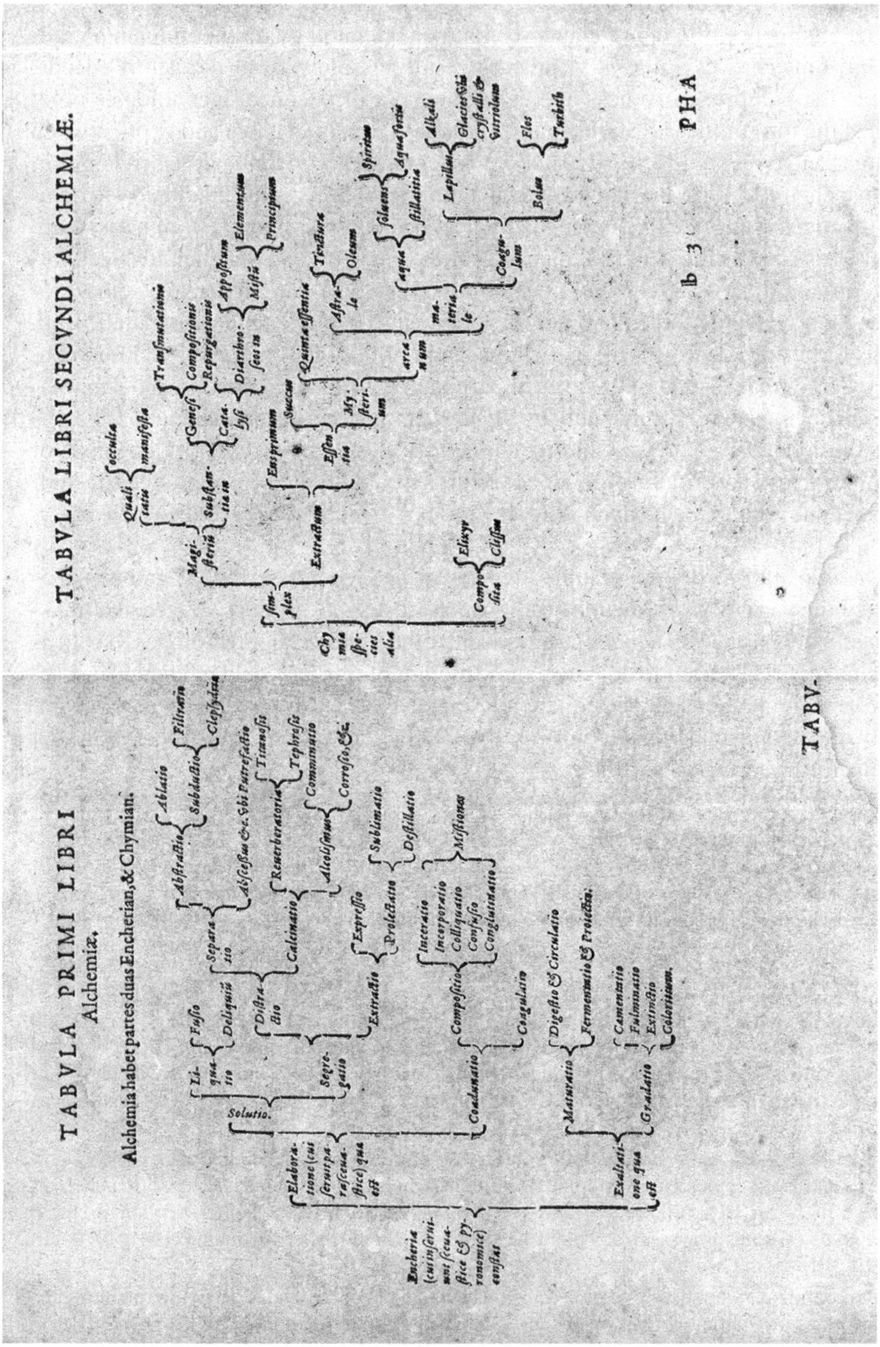

Fig. 3.2 Libavius' division diagrams of the Alchemy (1606)
Andreas Libavius: *Alchymia*, tabulae primi et secundi libri

(expression) as kinds of *Extractio* (extraction). Relationships of the various categories of substances then in use are rendered likewise, e.g. – again from Libavius' diagrams – *Quinta essentia* and *Arcanum* are classified as the two kinds of *Mysterium*, which, together with *Succos* (moisture), is listed as one kind of *Essentia*, and so on.

3.5 Terms and Concepts

Turning now to the explicit or implicit concepts held by sixteenth and early seventeenth century authors of chemical literature, the very names of the categories used in Libavius' and Beguin's diagrams merit attention. Libavius and Beguin did not invent or introduce a new nomenclature but used an age-old language in common use in this technological field, notwithstanding differences among authors about the precise meaning of certain of these terms.[34] We can leave open the question of whether this nomenclature was equally (and consistently) used in all branches of the field, and can also ignore the question of whether all practitioners or only educated ones used and understood these names. (The latter question concerns not only Latin names but technical terms in the various vernaculars as well and, thus, the social fabric of this field's practical knowledge.) What is of interest here is the question: How theory-laden is this nomenclature?

At first glance, many names of the several operations appear quite innocent in this respect, e.g. separation, extraction, smelting, dissolving, etc. On closer inspection, however, seemingly neutral names, e.g. distillation and sublimation, appear indicative of an understanding of the processes at hand that sheds a strange light on all of these technical terms. As one reads in Brunschwig's *Kleines Destillierbuch* (1500)

> *Distilling is nothing else than separating the subtle from the coarse /and the coarse from the subtle/making the instable or corrodible indestructible the material immaterial/the bodily more spiritual maintaining the incorporeal so that the spirit due to its subtlety may so much easier and faster penetrate the corporeal with his virtues and force [...]*[35]

And in Crollius' *Basilica chymica* (1609) one can read:

> *[...] For our Remedies require preparations, separations and exaltations before they can impact their hidden and restrained vertues. [...] They must be freed] from those scurvy raggs wherin they were wrapt up by a due separation from impurities and corruptible and filthy mixture of superficiall and externall Elements that pure and Christiline matter may be administered to our bodies.*[36]

---

[34] See for instance the entries "Extraction" or "Quinta essential" in the glossary of Klein (1994).

[35] "Diſtillieren nichtz anders iſt dan das ſubtyl von dem groben / vnd das grob von dem ſubtilen zů ſcheiden / das gebrechlich oder zerſtörlich vnzerſtörlicher zů machen das materialiſch vnmaterialiſcher zů machen / das lyplich geiſtlicher zemachen / das vnlieplich lieplicher zů behalten/ vff dz. lieplich der geiſt dz. lyplich durch ſin ſubtilithet deſter lichter dar zů behender dringen vnd penetrieren mag mit ſiner tugende vnd krafft [...]."Brunschwig (1500) f. 90r/v.

[36] For the English translation, see Henry Pinell: *Philosophy Reformed & Improved in Four Profound Tractates: The I. Discovering the Great and Deep Mysteries of Nature: by that Learned Chymist & Physitian Osw: Crollius* .... London: M.S. for Lodowick Lloyd, 1657, p. 93 f.

Two points are of interest here.[37] *First*, the intimate blending of a technical and a kind of natural philosophical language. Separating the coarse parts of a substance from the subtle ones means separating the bodily, material, and corrodible from the spiritual, immaterial, and indestructible. Clearly, this understanding of separating by distilling is inspired by the process taking place inside the distillation apparatus; but no less clearly, it associates understandings with this that go far beyond the technical operation and cannot be derived from it. That is to say, a technical process like distilling may have been seen as a model or instance of a philosophical, ontological or, as Multhauf put it, "ideological" concept of matter, a philosophical understanding, however, that did not originate in this technical sphere. This holds not only for operational terms such as separating, extracting etc. but also, and even more manifestly, for the designations of the various products of these chemical operations – *Spirit*, *Essence*, *Arcanum*, *Quintessence* etc.

Second, reading the quotation from Crollius's work alone, one might assume that this understanding of separating by distilling etc. is characteristic of beliefs held by followers of Paracelsus. However, as shown by the quotation from Brunschwig, who represents pharmaceutical chemistry before Paracelsus, this understanding of matter was obviously of a very general nature and entailed assumptions that underlay various specific chemical philosophies – the Aristotelian philosophy of elements (*Earth*, *Water*, *Air*, and *Fire*) along with the concept of a *Substantial form* and of a *Mixtum* as well as the early modern Paracelsian philosophy of *Elements* and *Principles* (*Sulphur*, *Mercury*, and *Salt*) and even the seventeenth-century natural philosophies of matter that assumed atoms and corpuscles as the ultimate entities that all substances consist of.

An Aristotelian philosophy of matter was the dominant one during the Middle Ages and must be seen as a common background of general theories linked to technical processes like distillation as indicated by Brunschwig and Crollius. Such theories disseminated aspects of Aristotle's philosophy of matter among chemical practitioners without any philosophical education.

Paracelsus' philosophy of *Elements* and *Principles* coincided with the Aristotelian philosophy in crucial assumptions, in particular in the understanding of the *Mixtum*.[38] Insofar as the deviations of Paracelsus' philosophy were not solely motivated by assimilations of other philosophical strands of the age but also by experiences and reflections of techniques and practices, it was the techniques and practices of Paracelsus as a physician, not as a chemist, as we will discuss shortly.

The same holds for chemists like Daniel Sennert (1572-1637), Sebastianus Basso (1573-?), Etienne de Clave (1587-1645) or later, and most prominently, Robert Boyle (1626-1692) who adopted an atomistic and corpuscular philosophy of matter. This must certainly be seen

[37] For the following, see also Multhauf (1961) 69.

[38] The Aristotelian *Mixtum* must not be mistaken for our present-day meaning of mixture as it meant a homogenous substance not consisting of but constituted by all four *Elements* or – in the case of Paracelsus – by all three *Principles*.

against the background of the contemporary revival of ancient atomism rather than as the result of any experiences in the sphere of chemical practices.[39]

3.6 Technology and Theory

With this, we encounter a very intricate ensemble of relationships between the technological and theoretical understandings characteristic of early modern chemistry. On the one hand, as we have seen, the understandings of the operations performed and processes employed typically have a dual nature: they are close to, or even inspired by technical processes, and at the same time they are expressions of a certain ontological understanding of matter. In other words, we find assumptions and beliefs about materials of a very general nature that are nevertheless modelled on technical processes – assumptions one could call general theories linked to certain technical processes. On the other hand, there are the overarching natural philosophical theories just mentioned (Aristotelian, Paracelsian, and atomistic and corpuscular philosophies) where such a link is hard to discern, or at least very dubious.[40]

In the case of Paracelsus' philosophy this appears less clear than in the other two cases. Such a link seems probable at first glance if we leave aside his version of the microcosm/macrocosm theory which interconnects planets, metals, and organs of the human body, and focus on his three principles *Sulphur*, *Mercury*, and *Salt*. For Paracelsus explicitly connected these principles with the process of combustion:

> *This piece of wood is a body. If you combust it, then that what burns is the Sulphur, the fume the Mercury, and that what becomes ashes the Salt. [...] One finds, thus, exactly three materials clearly distinguished. Each body disintegrates into these three materials. And in case that they don't appear clearly to the eye, artificial methods exist that render them discernible. That what burns is Sulphur and only Sulphur burns. That what fumes is Mercury and nothing else than Mercury sublimates (volatilizes), and what becomes ashes is Salt and nothing else than Salt can become ashes.*[41]

In light of this statement, it probably goes without saying that Paracelsus' principles cannot refer to our elements sulfur and mercury or to the compounds we classify as salts. And as regards the properties attributed to the three principles by Paracelsus, it becomes obvious

---

[39] For Aristotle's philosophy of elements, see book II of his *De generatione et corruptione*; see also Horne (1966). For the philosophy of Paracelsus and the Paracelsians, see Debus (1977) and Klein (1994) 38ff. For the revival of ancient atomistic philosophy at the turn of the sixteenth century, see Laßwitz (1890) vol. I, book 2, sect. 6–8 and Klein (1994) 57ff. – Robert Boyle, when reflecting the pros and cons of these overarching chemical philosophies in his *The Sceptical Chymist* (1661) actually discussed a plethora of chemical operations and findings supporting or contradicting each of these philosophies. See Klein (1994) 56ff. But this does not mean that such operations and findings inspired his corpuscular philosophical understanding of matter, nor that his corpuscular speculations were useful for a better understanding of such operations and findings. See Chalmers (2010).

[40] Regarding early modern atomistic and corpuscular philosophies of matter, see e.g., Meinel (1988).

[41] Paracelsus *Opus paramirum*, book I, chap. 2. In: Paracelsus (1926–1932–1932) I 68; translation W.L.

that they cannot be inspired, let alone derived, from the process of combusting. As we read in his treatise *Über die Entstehung der Metalle und Minerale*:

> *Notice that all colors derive from the Salt since the colors, the Balm and the coagulation [sc. of all materials] is provided by the Salt. The corporeality, substance, and structure [sc. of all materials] is provided by the Sulfur. Their properties, virtues, and Arcana by the Mercury.*[42]

Intricate relationships can be observed between the two kinds of general philosophical understandings of matter distinguished here – assumptions or theories with or without a link to the technical processes of contemporary chemical practices –, those of mutual presuppositions alongside those of indifference, since the assumptions with such a link proved to be compatible with each of the philosophies without such a link. We will not examine these relationships further. Our main question concerns whether, how and to which extent the developments of technological and theoretical texts in the field of early modern chemistry depended on each other. We should state that, up to the second half of the seventeenth century, natural philosophical theories of substances chemists dealt with on the one hand, and descriptions and reflections of the procedures and processes with these substances, on the other, had a life of their own. Before the second half of the seventeenth century, changes in the philosophical realm did not result in new or altered technological descriptions, instructions, or reflections, let alone in new practices; nor did changes in the technological realm result in new or altered theoretical assumptions. This striking finding becomes apparent in the famous series of manuals published by seventeenth-century French iatrochemists,[43] namely by the structure of these books. All of them[44] start with a short section (invariably making up less than a tenth of the manual) with theoretical statements confirming the iatrochemists' basics; next comes another very short section (based on the example of Libavius' *Alchemia*) containing descriptions and explanations of the various techniques and tools as well as technical terms used by chemists. This is followed by lengthy parts that discuss, on the one hand, operations with mineral substances such as metals and, on the other, provide detailed and elaborate explanations and instructions about how the various items of the huge realm of known and endorsed remedies and other commodities are manufactured. But – and this is the crucial point here – as a rule these explanations and instructions never refer back to the theoretical first section.

[42] Paracelsus: *Das Buch über die Minerale*. In: Paracelsus (1926–1932–1932) III 1041; translation W.L.

[43] Jean Beguin: *Les élémens de chymie* (see note 32 above), Étienne de Clave: *Le Cours de chimie*, (Paris: Olivier de Varennes, 1646), Nicolas Le Fèvre: *Traicté de la Chymie* (Paris. T. Jolly, 1660), Christopher Glaser: *Traicté de la Chymie* (Paris: by the author, 1663), and Nicolas Lemery: *Cours de chymie* (Paris: by the author, 1675).

[44] For the following, see Klein (1994) 119ff.

3.7 Unnoticed Shifts in Technology

These manuals are interesting in view of the technological developments in the chemical practices of the seventeenth century, especially those in pharmacy. They show impressively how far chemical medicaments in general and those produced from mineral substances in particular had become predominant in the pharmaceutical craft since the days of Paracelsus. The manuals' tables of content already show how, step by step, operations with mineral substances outnumbered the traditional ones with vegetable and animal materials – outnumbering but not replacing the latter.

> Beguin's *Élémens de Chymie* (1620), book II, deals with chemical operations, containing ca. 100 pages on distillation of vegetables but also of various salts for gaining mineral acids, followed by ca. 80 pages on calcination of metals (particularly antimony and mercury), and finally ca. 70 pages on salts. In de Clave's *Cours de Chimie* (1646), medicaments prepared from vegetable and animal materials (including turpentine and ammoniac) are dealt with in around 50 pages in book II whereas the roughly 100 pages of book III are dedicated to preparations from mineral materials – salts, acids, metals – and particularly to dissolutions of metals in acids. Le Fèvre's *Traicté de la Chymie* (1660), the most extensive of these manuals, contains preparations from animal and vegetable materials dealt with in book II (ca. 80 and ca. 200 pages respectively). Book III contains descriptions on further vegetable substances such as oils, resins etc. (ca. 150 pages), while processes with mineral materials – Earths, stones, metals, salts, acids, alkalis – occupy another 500 pages. In book II of Christopher Glaser's *Traicté de la Chymie* (1663), descriptions of and instructions on operations with metals, salts, acids, etc. occupy ca. 200 pages whereas descriptions of those with vegetable materials fill ca. 100 pages and those with animal materials ca. 25 pages. In Lemery's *Cours de Chimie* (1675) finally, ca. 330 pages are assigned to operations with mineral substances, ca. 150 to those with vegetable materials, and ca. 30 to those with animal materials.

At first glance, what seems to have taken place since the sixteenth century are merely quantitative shifts regarding the status and importance of the three classes of materials – from the animal, vegetable, and mineral domains. In principle, instances of nearly all of the operations – not only in the field of vegetable and animal materials but also in the increasingly dominant field of minerals – can be traced back to the sixteenth century and even to before Paracelsus' time. This also holds for the production of mineral acids by distilling natural salts[45] and for the dissolution of metals in such acids – two operations of particular significance for the development of chemical technology at the time. Furthermore, and again at first glance, this quantitative shift regarding the materials processed seems to have been accompanied only by refinements and sophisticated improvements to known chemical procedures – separation, dissolution, distillation, extraction, and so on – not by inventions and introduction of new ones.

---

[45] Nitric acid (*aqua fortis*) had already been gained from saltpeter in the Middle Ages. Methods of gaining sulfuric acid from alum or other natural sulfates were known in the Arabic Middle Ages. Only the production of hydrochloric acid from common salt is an achievement of the seventeenth century; it is attributed to Basil Valentine.

On second thoughts, however, the significance of this shift in the weight of operations with mineral materials becomes clear. This implied namely a shift in the respective weight of the two kinds of natural potencies utilized for processing materials by chemists: physical powers on the one hand and chemical powers on the other. To avoid misunderstandings, this requires a short explanation: "Physical powers" or agents refers first of all to heating[46] but also to mechanical potencies for separating, shredding, etc., that is, to thermic or mechanical processes that may or may not induce chemical processes. "Chemical powers" means solely chemical interactions, that is, chemical reactions between pure chemical substances.[47] The utilization of physical powers in this sense had been dominant for as long as vegetable and animal materials constituted the bulk of the substances processed. The utilization of chemical powers in this sense first gained prominence with the extension of operations with mineral materials which – as in the case of metals or salts – either consisted of or contained such pure chemical substances.

The historical actors did not see this shift in the way we do in hindsight. For us, they were processing mixtures by their operations with vegetable and animal materials and the result of the operations was the separation of the various substances mixed into these materials. These separations must not be mistaken for chemical analyses where some or all components of a chemical compound are separated by chemical reactions. True, when these vegetable or animal materials are subjected to distillation, cascades of chemical reactions will take place – reactions, however, that would be a serious challenge even for modern chemists to recognize or reconstruct. It therefore comes as no surprise that these chemical reactions did not exist for the historical actors who only realized the fire as the decisive agent of this process of separating or – in their terminology – of *exalting* the medically valuable substances.

> *[…] For our Remedies require preparations, separations and exaltations before they can impact their hidden and restrained vertues. As all things are proved by Fire, so also the Tryal of the knowledge of Physick [scientia medicina] is to be made by Fire: Physick [medicina] and Chymistry [chymia] cannot be separated. For Chymistry […] does make manifest not only the true Simples, Wonders [Magnalia] Secrets [Arcana], Mysteries [Mysteria] Vertues, Forces respecting health, but also in imitation of the Archaeon Ventricle or Natural In-bred Chymist, it teacheth to segregate every mystery into its own reservacle [reservaculum], and to free the medicines from those scurvy raggs wherin they were wrapt up by a due separation from impurities and corruptible and filthy mixtrure of superficiall and externall Elements that that pure and Christiline matter may be administered to our bodies. But to deliver this from prison and captivity, Hoc opus, hic labor est, is a hard task to performe.*[48]

This statement by Oswald Crollius of 1609 was by no means outdated or disputed when Estienne de Clave, Nicolas Le Fèvre, and Christopher Glaser published their manuals in the mid-seventeenth century.

[46] See, for instance, Debus (1967) and Holmes (1971).

[47] The term *pure chemical substances* is not, of course, an actors' term; it is used for designating either chemical elements or chemical compounds (in the present meaning of the terms).

[48] Oswald Crollius: *Basilica Chymica*; for the English translation see Henry Pinell (note 36 above); for the Latin terms of the original added in brackets, see *Basilica Chymica* (edition Frankfurt, ca. 1611), reprint Hildesheim: Olms, 1996, 48 f.

Not realizing that different processes were at work when the same operation – say, a dissolution or a distillation – was performed with vegetable or mineral materials, the historical actors understood the latter in the same way as the former. The dissolutions of metals in mineral acids, for example, was not even seen as a true chemical alteration – that is, as an exaltation of a substance by freeing its restrained virtues – let alone as a chemical reaction of the two substances that led to the chemical synthesis of the metal with a component of the acid. Rather, it was assumed to be the same separation process by which organic mixtures were prepared for distillation by processes of a more mechanical nature like washing out or crushing and grinding. Accordingly, the dissolution of metals in acids was named *Calcination* or *Cementation*, that is, it was designated as a process by which the metal was pulverized.

3.8 Irritating Processes and New Understandings

These traditional understandings of chemical operations did not, however, remain unshaken in the course of the seventeenth century. In particular, certain operations with mineral materials caused irritations which could not be ignored in the long run. One of these puzzling operations was the production of salts by dissolving mineral substances such as metals in acids with a subsequent precipitation, that is, by adding a further substance that caused the desired salt to fall out of the solution. Equally irritating were operations by which dissolved or apparently consumed substances could be regained, that is, the case of reversible chemical processes. Such reversible processes occurred mainly in metallurgy, although also in connection with consumed acids that could regained under certain circumstances.

The utilization of reversible chemical processes for refining of precious metals was a standard practice of smelters and assayers that can be traced back to the Middle Ages or even to Antiquity. The extraction of *Silver* out of copper-silver alloys, for example, was performed by the so-called *Saiger process*. In this process, the copper-silver alloy was fused with lead whereby, along with almost pure copper, a silver-lead alloy resulted out of which the silver was extracted by cupellation, that is, by high-temperature smelting. Thus, the lead applied is regained in the end. *Gold*, to give another example, was extracted out of several of its alloys by amalgamation, that is, by adding mercury to these alloys. Out of the resulting gold-mercury alloy, the gold and the mercury could be separated by distilling off the mercury. Again, the substance applied, the mercury, is regained in the end.

Reversible processes such as those utilized by the *Saiger process*, as well as salt production by dissolution in acids with subsequent precipitation represented, indeed, a challenge for the then prevailing understanding of chemical operations on the basis of either an Aristotelian or a Paracelsian and, by the way, no less an atomistic philosophy of matter.

We need not go into detail about the irritations these operations caused. It may suffice to point to some questions that must have puzzled the historical actors. For instance, in the case of salt production by acid dissolutions and subsequent precipitation, how could a substance with medically valuable virtues be created by art without the help of fire? And how did the added substance effect the falling out of the salt? Or, in the case of reversible operations, why did silver alloy with lead more easily than gold? And what happened to retrievable substances during the operations by which they had been consumed: Did they remain unaltered in spite of the action of fire? – and if so, where and in which form did they remain; or were they *mortified* and *resurrected*, to put it in alchemical language?

Metallurgy was the first domain where these puzzling operations were accompanied by understandings that were latently incompatible with the prevailing Aristotelian or Paracelsian philosophies of matter. Such understandings were already expressed in publications by metallurgists in the sixteenth century. Regarding the separation of silver from a silver-tin alloy by adding lead, we can read in Lazarus Ercker's *Aula subterranea* (1573):

Among the metals, only tin goes very readily into lead. But it does not suffer the fire's power. For, as soon as it is strongly heated it wants to go away, rises out of the lead and [...][49]

Likewise, regarding the amalgamation process, we find in Agricola's *De re metallica* (1556) the understanding that at a certain phase of the process the mercury "has absorbed the whole gold."[50] In the same vein, metallurgists, in contrast to pharmaceutical chemists, understood the dissolution of metals in acids as an incorporation of the metal into the acid. That is, the dissolved metals were assumed to be preserved in dissolutions – lead alloys, amalgams, or nitric acids – in spite of the homogeneous appearance of these dissolutions. It seems very likely that the experience and utilization of reversible processes played a decisive role for the development of these understandings. As late as in the mid-seventeenth century, Johann Rudolph Glauber who rejected these understandings of metallurgists, voiced exactly this assumption:

Since the majority of smelters that use the Saiger process do not understand how this process works and assume that the corporal lead runs, along with the copper and other admixtures, into the cupels because of the possibility to regain it by smelting [...][51]

With hindsight we can take these understandings as symptoms and the first conscious reflexes of the emerging domain of chemical practices with pure chemical substances that was developing alongside the traditional and still dominant domain of chemical practices that processed mixtures. However awkward these conscious reflexes may appear, they indicate, or at least imply, essentials of a chemistry of

[49] "Das Zinn ist unter den anderen Metalln allein / das sehr gern ins Bley gehet / aber in der gewalt des Fewers wil es nichts darbey leiden / dann so bald grosse hitz darzu kompt / so wil es wider davon und steigt aussm Bley auff / wird [...]" Ercker: *Aula subterranea* (1573) f. 29v.

[50] "[...] deinde in eandem patinam infunditur frigida, ac mox argentum uiuum, quod aurum omne abſorbuit, à reliquo ramento lotura collecto ſecretum in unum confluit [...]" Agricola *De re metallica* (1556) book 7, 209.

[51] "Dan der mehrer Teil die mit Abtreiben [Metallscheiden mit Blei] umgehen / nicht wissen wie es hergehe mit dem abtreiben / sondern vermeinen das Bley lauffe also corporalisch mit dem Kupfer und anderm Zusatz in die Capellen / weiln es sich wider corporalisch darauß schmeltzen lässt [...]"Glauber *Furni novi philosophici* (1646–48) IV 14. Glauber himself assumed that the lead was transformed into a slag by the Saiger process; see ibid.

pure chemical substances, namely that chemical substances react with one another, enter into as well as leave conjunctions, are preserved in such conjunctions, show preferences regarding conjunction partners, and, last but not least, that all of these reacting, conjoining and dissociating substances are of the same ontological rank.

However, not only smelters and assayers, but iatrochemists, too, had to deal conceptually with chemical practices that utilized interactions of pure chemical substances. As already seen, the production of salts had become a major occupation in their own pharmaceutical field during the seventeenth century. Accordingly, salt production by dissolution of metals and other minerals in acids with subsequent precipitation by addition of a third substance became a major topic of reflections in technological writings of the last third of that century.[52]

In the course of these reflections,[53] the meaning of several notions pertaining to the production of salts was slightly or profoundly changed. This concerned even the notion "salt" itself, which only then became the designation of a genus of chemical substances comprising several species of salts – natural and artificial salts, e.g. vitriols found in nature and vitriols produced in the laboratory, *sels moyens* (middle salts), *sels salés* or *salia composita*, and so on. Without going into detail here, we will point to one principal feature of the latter terms, namely that they represent a classification of substances according not to the procedure employed for their production or their virtues, but to their composition. Another noteworthy change concerned the notion "alkalis," which came to mean bases of salts since, beside the traditional two alkalis, the fixed and the volatile one, it now also comprised metals and earths, that is, the two other kinds of substances that are dissolved in acids in the process of salt production. Perhaps even more important than these changes was the emergence of a new understanding of the notion "dissolution" or "coagulated dissolution" which came to designate a conjunction of chemical substances that remain preserved in this conjunction. This notion of a "coagulated dissolution," developed by pharmaceutical chemists, coincided in a crucial way with the metallurgists' broad concept of dissolutions – alloys, amalgams, nitric acids – in which a metal is dissolved and preserved notwithstanding the homogeneous appearance of these dissolutions. This common core of the two concepts – preservation in dissolution – constituted the bridge that allowed the conceptual unification of the chemical practices with pure chemical substances in metallurgy and in pharmaceutical chemistry. It seems appropriate to add that this common core of the concept of dissolution coincides to a considerable extent with essentials of the modern notion "chemical compound."

3.9 A New Theory

As we are not writing a history of early modern chemistry, we will not continue tracing the steps by which these beginnings developed into a new theory that is a chemical theory in the modern meaning of the term, namely, a theory of laws

[52] These reflections can be found in particular in Lemery's *Cours de chymie* (1675) and in several of Wilhelm Homberg's articles published by the *Académie Royale des Sciences Paris* in the first decade of the eighteenth century, especially "Essays de Chimie" (1702) and "Memoire. Touchant les Acides & les Alcalis, pour server d'addition à article du Sel principe" (1708).

[53] For the following, see Klein (1994) 227ff; Holmes (1989) 33ff.

governing the behavior and interactions of chemical substances. Rather, pursuing our investigations of the interplay or co-evolution of technological and scientific literature for the field of chemistry in this period, it is of particular interest to determine what this new theory owed to the technological literature and conversely, the role of this new theory in further developments of technological investigations. For this we will focus on features of this interplay as they appear in one key document of the new chemical theory, namely Etienne François Geoffroy's (1672–1731) *Table des differentes rapports observés en Chimie entre differentes substances* of 1718. (Fig. 3.3).[54]

Fig. 3.3 E.F. Geoffroy's affinity table (1718)
Etienne François Geoffroy: *Table des differentes rapports [...].* – Klein (1994), p. 21

[54] *Histoire de l' Académie Royale des Sciences,* Paris: Imprimerie royal, 1718, p. 202–212. An English translation by Andrew Mendelsohn was published in *Science Context* IX/3 (1996) 313–319. Etienne François Geoffroy, the son of an apothecary, was an apothecary's apprentice himself before studying medicine; he became professor of chemistry at the *Jardin des Plantes*, Paris, then of chemistry and medicine at the *Collège Royal* and finally of medicine at the *Faculté de Médicine* in Paris.

To give just a quick orientation[55]: Geoffroy's *Table* is a paper submitted to and published by the French academy of science which includes a table explained by a short text. The table classifies chemical compounds formed in the processes of salt production as well as in the processes of metal separation in metallurgy. The compounds are arranged in columns. Each column (except one) stands for a class of compounds formed by the substance at the top of the column, the reference substance, and the substances that can be combined with it. Of the table's 14 horizontally juxtaposed columns, 13 present such classes – 7 the classes of compounds formed in acid solutions, 6 the classes of those formed in metallurgical processes – and 1 column shows products of watery solutions. The table indicates furthermore the affinity (*disposition à s'unir*) between these substances and the reference substance. The respective degree of affinity of a substance in a column is indicated by the substance's position in the vertical sequence of substances beneath the reference substance. By distributing degrees of affinity to the substances of a class, the table allows the reader to deduce which substance can dissolve another one out of its conjunction with the reference substance by combining itself with the latter. That is to say, the table shows the order of possible replacement reactions in each class of compounds.

Clearly, Geoffroy's *Table* is not an instance of technological literature. Rather, it belongs to theoretical literature, structuring and classifying the then known chemical processes with pure chemical substances, thereby integrating and transforming them into elements of a connected and coherent field of chemistry. Moreover, it also demonstrates rules that govern the behavior of these substances. Geoffroy's *Table* is rightly appreciated as a key document of the incipient chemical theory of pure chemical substances from which our modern-day science of chemistry developed.

It goes without saying that Geoffroy could accomplish this integration and arrangement of the contemporary chemistry with pure chemical substances only on the basis of the technological chemical literature of the sixteenth and seventeenth centuries which provided him with all information regarding the state of chemical art.[56] He was helped by the reflections and theoretical assumptions dispersed in this literature and particularly, as we have seen, in the literature of the last third of the seventeenth century. In contrast to the short theoretical parts at the beginning of the seventeenth-century *Cours de chimie*, Geoffroy's theory is not an attempt at subsuming chemical processes under an Aristotelian or other philosophy of matter that is presupposed and had been developed long ago, independently of any knowledge of these processes. Rather, his theory is a theory of patterns and laws of precisely the processes the technological literature of the time describes. His theory is closely associated with this literature.

As regards the impact of his theory on the subsequent technological chemical literature, one has to distinguish between the two main domains of chemistry – the chemistry that processed mixtures of the vegetable and animal kingdoms and the chemistry with pure chemical substances – at the time almost only mineral (or inorganic) substances. For the latter, Geoffroy's *Table* was certainly an important

[55] For a detailed description and interpretation of Geoffroy's *Table*, see Klein (1994) part II/1 and VI/3 as well as Klein (1995). See also Holmes (1996).

[56] Ursula Klein has shown that each of the processes forming compounds that are explicitly or implicitly enlisted in Geoffroy's *Table* can be traced to descriptions in the technological literature – see Klein (1994) Tables 1 and 2.

reference point as proved by the series of affinity tables which were composed and published in the course of the eighteenth century.[57] For the former, Geoffroy's theory naturally had almost no significance. The ways in which eighteenth-century chemists tried to bring order into this huge domain of vegetable and animal materials were of a completely different character.[58] It should be added that this chemical domain was then not simply much larger than that of pure chemical substances but also of greater practical relevance in several respects – medical, nutritional, hygienic, etc. – and, not least, economically.

3.10 Conclusion

In conclusion, we should first state that in the field of early modern chemistry before the eighteenth century, technological and scientific literature are mostly indistinguishable; that is, casual statements on or deliberate and comprehensive expositions of the theoretical assumptions held by the author were part and parcel of almost every significant technological work on chemistry since Brunschwig's *Destillierbuch* of 1500. In the seventeenth century it became the standard form of chemical manuals to begin with a theoretical part before providing the various technical descriptions and instructions. A close interplay between developments on the technological and theoretical level could be expected under these circumstances.

However, as we have seen, these two levels had very little to do with each other. True, theoretical terms like extraction, sublimation or distillation were referring to technical processes but at the same time they conceptualized these processes in the framework of natural philosophical theories of the ultimate constitution of matter that originated and developed completely independently of chemical technologies. This also holds for the succession of early modern theories of the ultimate constitution of matter – from more or less Aristotelian to Paracelsian ones and furthermore to atomistic matter theories of the seventeenth century. And just as these philosophical theories of the ultimate constitution of matter owed nothing to developments of chemical procedures or the expansion of substances processed by them, conversely, these technological developments owed nothing to these theories.

Finally, we encountered the beginnings of a new theory, emerging around 1700, that was not yet another philosophical matter theory but a true chemical theory, that is, a theory of laws that rule the behavior and interactions of chemical substances. This theory resulted from reflections of metallurgical processes and, above all, of new procedures for salt production that spread in the seventeenth century – reflections on the basis of a rich technological literature on these processes and

---

[57] For eighteenth-century affinity tables, see Duncan (1970) and (1996). These tables have a double character: They must be regarded as technological literature as all of them provide useful information about possible replacement reactions for practitioners; some of them are also of a theoretical nature as they include rearrangements of the field of chemistry with pure chemical substances.

[58] For these ways, see Klein and Lefèvre (2007) part I.

procedures. The close relationship of technological and scientific literature is rarely as clear as in this case, where a scientific theory developed by studying and conceptualizing natural laws as utilized in technological processes and described and documented in technological manuals.

References

Anderson, Frank J. 1977. *An Illustrated History of the Herbals*. New York: Columbia University Press.

Arber, Agnes. 1912. *Herbals, their Origin and Evolution: 1470–1670*. Cambridge: Cambridge University Press.

Bernardoni, Andrea. 2011. *La Conoscenza del fare: Ingegneria, arte, scienza nel De la pirotecnia di Vannoccio Biringuccio*. Roma: L'Erma di Bretschneider.

Biringuccio, Vannoccio. 1959. *The Pirotechnia of Vannoccio Biringuccio*. New York: Basic Books.

Brunschwig, Hieronymus. 1500. *Liber de arte distillandi de Simplicibus*. Straßburg: Hans Grüninger.

Cennini, Cennino. 1899. *The Book of the Art of Cennino Cennini; a Contemporary Practical Treatise on Quattrocento Painting*, trans. Jane P. Christiana Herrington. London: G. Allen & Unwin.

Chalmers, Alan F. 2010. Boyle and the Origins of Modern Chemistry: Newman Tried in the Fire. *Studies in History and Philosophy of Science* 41: 1–10.

Clarke, Mark. 2001. *The Art of All Colours: Mediaeval Recipe Books for Painters and Illuminators*. London: Archetype Publications.

Debus, Allen G. 1967. Fire Analysis and the Elements in the Sixteenth and Seventeenth Centuries. *Annals of Science* 23: 127–147.

———. 1977. *The Chemical Philosophy: Paracelsian Science and Medicine in Sixteenth and Seventeenth Centuries*. New York: Science History Publications.

Duncan, Alstair M. 1970. The Function of Affinity Tables and Lavoisier's List of Elements. *Ambix* 17 (1): 28–42.

———. 1996. *Laws and Order in Eighteenth Century Chemistry*. Oxford: Clarendon Press.

Eamon, William. 1994. *Science and the Secrets of Nature: Books of Secrets in Medieval and Early Modern Culture*. Princeton University Press: Princeton.

Fischer, Hans. 1966. *Conrad Gessner: Leben und Werk*. Zürich: Kommissionsverlag.

Forbes, Robert James. 1948. *Short History of the Art of Distillation*. Leiden: Brill.

Ganzenmüller, Wilhelm. 1941. Liber Florum Geberti. Alchemistische Öfen und Geräte in einer Handschrift des 15. Jahrhunderts. In *Quellen und Studien zur Geschichte der Naturwissenschaften und der Medizin*, ed. WIlhelm Ganzenmüller, 273–304. Berlin: Springer.

Holmes, Frederic L. 1971. Analysis by Fire and Solvent Extractions. The Metamorphosis of a Tradition. *Isis* 62 (2): 129–148.

———. 1989. *Eighteenth-Century Chemistry as an Investigative Enterprise*. Berkeley: University of California.

———. 1996. The Communal Context of Etienne François Geoffroy's 'Table des rapports'. *Science in Context* 9 (3): 289–311.

Horne, R.A. 1966. Aristotelian Chemistry. *Chymia – Annual Studies in the History of Chemistry* 11: 21–27.

Jaumann, Herbert. 2004. *Handbuch Gelehrtenkultur der Frühen Neuzeit*. 2 vols. Berlin: Walter de Gruyter.

Klein, Ursula. 1994. *Verbindung und Affinität. Die Grundlegung der neuzeitlichen Chemie an der Wende vom 17. zum 18. Jahrhundert*. Basel: Birkhäuser.

———. 1995. E.F. Geoffroy's Table of Different 'Rapports' Observed between Different Chemical Substances. A Reinterpretation. *Ambix* 42 (2): 79–100.

Klein, Ursula, and Wolfgang Lefèvre. 2007. *Materials in Eighteenth-Century Science. A Historical Ontology*. Cambridge, MA: The MIT Press.
Laßwitz, Kurd. 1890. *Geschichte der Atomistik vom Mittelalter bis Newton*. Hamburg: Voss.
Laurie, Arthur P. 1910. *The Materials of the Painter's Craft in Europe and Egypt, from the Earliest Times to the End of the XVIIth Century*. London: Foulis.
Long, Pamela O. 2001. *Openness, Secrecy, Authorship: Technical Arts and the Culture of Knowledge from Antiquity to the Renaissance*. Baltimore: John Hopkins University Press.
Meinel, Christoph. 1988. Early Seventeenth-Century Atomism: Theory, Epistemology, and the Insufficiency of Experiments. *Isis* 79: 68–103.
Merrifield, Mary P. 1999. *Medieval and Renaissance Treatises on the Art of Painting*. New York: Dover.
Moran, Bruce T. 1991. *Chemical Pharmacy Enters the University: Johannes Hartmann and the Didactic Care of Chymiatria in the Early Seventeenth Century*. Madison: American Institute of the History of Pharmacy.
Moran, Burce T. 2007. *Andreas Libavius and the Transformation of Alchemy: Separating Chemical Cultures with Polemical Fire*. Sagamore Beach: Watson Publishing.
Multhauf, Robert P. 1961. Libavius and Beguin. In *Great Chemists*, ed. Eduard Farber, 65–79. New York: Interscience Publishers.
Olschki, Leonardo. 1965. *Geschichte der neusprachlichen wissenschaftlichen Literatur*. Reprint ed., 3 vols. Vaduz: Kraus.
Paracelsus. 1926–1932. *Sämtliche Werke nach der zehnbändigen Huserschen Gesamtausgabe (1589–1591)*. Jena: Gustav Fischer.
Partington, J. R. 1961–1970. *A History of Chemistry*. 3 vols. London: Macmillan.
Rex, Hannelore. 2001. *Die Lateinische Agrarliteratur: Von den Anfängen bis zur frühen Neuzeit*. Wuppertal: Universitätsbibliothek.
Sisco, Anneliese Grünhaldt, and Cyril Stanley Smith. 1949. *Bergwerk- und Probierbüchlein*. New York: The American Institute of Mining and Metallurgical Engineers.
———. 1951. *Lazarus Ercker's Treatise on Ores and Assaying: Translated from the German Edition of 1580*, trans. Cyril Stanley Smith and Martha Teach Gnudi. Chicago: The University of Chicago Press.
Underwood, E.A. 1972. Franciscus Sylvius and His Iatro-Chemical School. *Endeavour* 31: 73–176.

Chapter 4
Gunnery

The subject of this chapter, gunnery, means not just the science of ballistics – internal ballistics, transition ballistics, external ballistics, terminal ballistics. Rather, in view of the interplay between technological and scientific literature of the early modern period, we will take gunnery as an essential or even pivotal part of the art of warfare in that period – the part which comprised, beside shooting, the fabrication of various kinds of guns, of gunpowder, projectiles, and the production of the means used for transporting, maneuvering, and deploying artillery as well as the practical field in which the artillery was mainly used at the time, that of attacking or defending fortified places.

As is well known, the early modern art of warfare grew up with the development of firearms in the late Middle Ages and the early modern period, and particularly with the co-evolution of artillery and fortification. These developments gave rise to new types of experts who procured, operated, maintained, and managed the new technical equipment of gunpowder and firearms warfare. To begin with it was blacksmiths and locksmiths who fabricated and operated guns; later on, when cannons were cast in decarburized iron and then in bronze, founding and operating guns became the tasks of different professionals – founders and gunners – although gunsmiths remained central figures for operating heavy artillery up to the 30 Years War.[1] The new demand for mechanical engineering expertise needed for devising gun mounts, carriages, hoists and other equipment shaped the figure of the early modern military engineer; and the figure of the military architect emerged through the developments of fortification. The development of these fields of specialized practical knowledge was accompanied by a growth in new technological literature and scientific literature on some aspects of gunnery.

In this chapter we will focus on the early modern literature on shooting, including the fabrication of the artillery (guns, gunpowder, and projectiles) as well as

[1] See, for instance, [Anonymus] (1757) part II; see also Damerow and Lefèvre (1985) 363.

W. Lefèvre, *Minerva meets Vulcan: Scientific and Technological Literature – 1450-1750*, Archimedes 60, https://doi.org/10.1007/978-3-030-73085-7_4

loading guns, firing, aiming, and determining the distance of a target by surveying. Fortification and issues of military strategies and tactics will be addressed only in passing where necessary.

4.1 Warfare in Illuminated Manuscripts

The practice of shooting and hurling projectiles by machines did not originate with firearms. Intricate devices such as the crossbow, bow-machine, counterweight catapult, ballista, trebuchet, etc. for launching projectiles such as arrows or darts, stones, incendiaries, etc. had been invented and applied in warfare long before the Middle Ages in several civilizations – not only those of the ancient Mediterranean and Near East. These mechanical throwing machines were not immediately ousted when firearms appeared on the battlefields. In the West, where firearms can be traced back to the thirteenth century, traditional machines remained in use alongside the gradually developing firearms well into the sixteenth century. It was by no means a retrogressive sign or a humanist obsession that ancient literature on warfare and weapons were still studied and translated in the fifteenth century.

Only a mere fraction of the literature on the art of warfare and on siege machinery from Greek, Hellenistic, and Roman Antiquity was transmitted to the West before the sixteenth century, namely the *Epitoma rei militaris* by Flavius Vegetius Renatus (fourth century AD) and book X of Vitruvius' *De architectura libri decem.* The former work was printed as well as translated into several vernaculars in the course of the fifteenth century. The translation into German by Ludwig Hohenwang (1440–1505?), printed in 1475, deserves attention in this context because it contains, as a kind of appendix, a collection of some 60 figures mostly of siege equipment, several of which depict "modern" firearms.[2]

The simultaneous use of traditional throwing machines and firearms in the fifteenth century is also documented in that century's illuminated manuscripts on machines, most of which prominently feature war machines. Guns, gun mounts and the like were pictured side by side with the traditional siege machinery but appear initially as an addition to the latter rather than as superior new kinds of weapons.

Among the c. 200 figures in Konrad Kyeser's (1366–1405) *Bellifortis* (c. 1405), one finds only three or five pictures of firearms, whereas several mechanical throwing devices are depicted alongside an abundance of other traditional siege devices (towers, grapplers,

[2] Most of the extant or known treatises of classical Antiquity on armaments, e.g. those by Ctesibius, Philo of Byzantium, Hero of Alexandria, Apollodorus of Damascus, Biton, or Athenaeus mechanicus, were not known at the time. Some of them were rediscovered in the sixteenth century, at a time when their content was only of interest to scholars of the ancient world. For this literature, see, for instance, Marsden (1969) and (1971). For Vitruvius' descriptions of war machinery, see *De architectura libri decem,* book X, Chaps. 10–17. Vegetius' *Epitoma* was printed (in Latin) in several places in the 1470s and 1480s. The German translation by Hohenwang was first printed in 1475/76 (Augsburg: Johann Wiener). For this edition, see Leng (2010) No. 39.21a. The figures in the appendix are copied from Roberto Valturio's *De re militaris libri XII* (1472).

battering rams, etc.). Giovanni Fontana's (1395–1455) *Bellicorum instrumentorum liber* (c. 1430) presented only traditional siege machinery, while Taccola (Mariano di Jacopo – 1381–1453), in the mid-fifteenth century, presented a number of guns, particularly bombards, alongside many traditional devices for carrying or throwing incendiaries. In Roberto Valturio's (1405–1475) *De re militari libri XII* (1466/1472) and in the *Anonymous of the Hussite Wars* (c. 1475) there are nearly equal proportions of pictures of mechanical throwing machines and guns/gun accessories. Finally, in Francesco Giorgio Martini's (1439–1501) *Arte Militare e Macchine Belliche Antiche e Moderne* (c. 1480), traditional and modern weapons are discussed side by side.[3]

In view of our question of the interplay between technological and scientific literature one might wonder whether or not these illustrated manuscripts and – as regards Valturio – printed books can be regarded as technological literature at all (see sect. 1 of the chapter on mechanical engineering.) In any case, it is almost always literature that presents existing weapons or proposals for weapons to princes or city state councils and their condottieri. The focus is on the use of these devices, not their technical details. The technique or "grammar" that the illustrations of these manuscripts and books use for transferring information has been summed up by Rainer Leng as follows:

> *All illustrations have to be self-explanatory. They must not presuppose a thorough technical understanding of the depicted devices. Such a condition would be misplaced in view of these manuscripts'/books' addressees. The focus must not lie on the technique of the devices but on their tactical application and the results that should be obtained by them.*[4]

The same holds to a greater or lesser extent for collections of such machine drawings compiled in the late fifteenth century as well as for illuminated inventories of arsenals which constituted a sort of arsenal on paper at courts or administrative boards of city states.[5] The emphasis of the drawings in these collections had been placed on the use, not the construction, of these devices. We should add, however,

---

[3]The numerous fifteenth-century copies of Kyeser's *Bellifortis* differ, among other things, as regards the number of illustrations, and therefore no exact statements can be made; see Leng (2010) No. 39.4. Fontana's *Bellicorum instrumentorum liber* presents some fantastic rocket devices which seem, however, to be unrelated to contemporary firearms. Taccola's *De Machinis* (München BSB, Codex Latinus Monacensis 28,800 II) is more informative in this respect than his *De Rebus Miliitaribus* (Paris BnF, Codex Latinus 7239). In his *De re militari* (Verona: Johannes Nicolai, 1472), Valturio depicted particularly large cast guns. In the *Anonymous of the Hussite Wars*, only "civilian" machines such as mills or hoists are accompanied by texts. For the *Anonymous of the Hussite Wars*, see B.S. Hall (1979), Leng (2002) I 231ff. Francesco di Giorgio Martini: *Arte Militare* – Giorgio Martini (1967) vol. I – (Codice L: Firenze, Bibl. Laurenziana, Ashburnham 361, f. 47v ff. and Codice T: Torino, Biblioteca Reale, Codice Saluzziano 148, f. 53v ff.)

[4]"*Das Grundprinzip lautet überall: Alle Illustrationen müssen selbsterklärend sein. Tiefere technische Hintergrundwissen darf nicht vorausgesetzt werden. Im Hinblick auf das Publikum ist dies auch gar nicht erwünscht. Nicht die Technik selbst steht im Vordergrund, sondern deren taktischer Einsatz und die zu erzielenden Resultate.*" Leng (2008) 48; see also Leng (2004).

[5]*Das mittelalterliche Hausbuch* (Fürstl. Waldburg-Wolfeggsche Bibl., ca 1480) is a collection of drawings, not a continuous illuminated manuscript. Of its ca. 20 folios depicting war machinery, 10 present guns/gun accessories, whereas only one mechanical throwing machine is represented. The *Weimarer Ingenieurkunst- und Wunderbuch* (Herzogin Anna Amalia-Bibl., Weimar, fol. 328, c. 1500) is a compilation of earlier sources. For the identification of most of its sources, see Leng

Fig. 4.1 *Karrenbüchse* (Gun Cart) (2nd Half of the fifteenth c)
Martin Merz: [*Feuerwerksbuch*] – BSB Cgm 599 – f. 13v

that the pictures of devices in some of these collections and inventories are so detailed and precise that they certainly transmitted technological knowledge to an inquiring viewer, let alone to an experienced mechanic. (Fig. 4.1) This may be due to the fact that one finds master gunsmiths (*Büchsenmeister*) among the responsible authors (not illustrators) of these collections, e.g. Martin Merz (1425–1501), Ulrich Bessnitzer (1450–1521), Philipp Mönch (?1475-?), and Franz Helm (ca. 1500–1567).[6] *Büchsenmeister* as authors deserve a closer look in our context.

4.2 Gunners' Manuscripts (*Büchsenmeisterbücher*)

Beside the aforementioned illuminated manuscripts, collections of drawings of war machinery and inventories of arsenals, another kind of literature on issues concerning the new firearms existed in the fifteenth century. It consisted of humble

(2002) II 292–296. For collections of drawings of armaments from around 1500, see Leng (2010) No. 39.8. For inventories of arsenals, see Jähns (1889) book III §§ 62 and 66–68.

[6]The *Büchsenmeisterbuch* of Martin Merz (Munich BSB Cgm 599, c. 1475); Ulrich Bessnitzer: *Landshuter Zeughausinventar* (Heidelberg, Cod. Pal. germ. 130, late fifteenth century); Philipp Mönch: *Dys büch der stryt vnd büchßen* (Heidelberg Cod. Pal. germ. 126, 1496); Franz Helm: *Buch von den probierten Künsten* (Heidelberg, Cod. Pal. germ. 128, 1535).

manuscripts without elaborate illustrations. These manuscripts, composed by practitioners, namely by *Büchsenmeister* for fellow *Büchsenmeister*, can be taken as early modern technological literature on gunnery in its initial stage.[7]

In the fourteenth and early fifteenth centuries, the field of duties of a *Büchsenmeister* was all-encompassing, comprising the fabrication of guns, gunpowder, and projectiles as well as, in military campaigns, the operation of guns and generally the repair and upkeep of one or more guns. In the course of the fifteenth century, due to the spread of the technique of manufacturing large guns (bombards) by casting, a specialization occurred between the producer and the user or operator of firearms, that is, between professional founders and the *Büchsenmeister* who remained responsible for the operation and upkeep of these weapons and often also for the arsenal of a city or a court.[8]

The eldest *Büchsenmeisterbücher* were written or composed by unknown *Büchsenmeister*. The core of these manuscripts in vernacular consisted of recipes for mixing different varieties of gunpowder. These recipes were sometimes combined with some simple drawings juxtaposed with short texts. Among these manuscripts, and possibly as a combination of them, one *Büchsenmeisterbuch* emerged that was to become a key example, namely the *Feuerwerksbuch* of 1420. It existed initially in many slightly different copies and versions before the 1440s, when it achieved the more or less stable core content which proved exemplary for succeeding texts of the genre.[9]

Because of this model status of the *Feuerwerksbuch*, it seems worthwhile to look briefly at its content. The manuscript begins, after a preface, with the twelve canonical gunners' questions concerning the operation of guns, particularly the effect of gunpowder and the fixing of the projectile in the gun barrel. A listing of the requisite abilities and skills of a competent gunner is followed by gunpowder recipes, particularly instructions for the acquisition and preparation (purification and processing) of its ingredients (saltpeter, sulfur, charcoal) as well as for reprocessing spoiled gunpowder. Instructions for loading guns and shooting, including precautions against explosions, conclude the standard content of the book, to which various final addenda, amendments, and the like may have been appended.[10]

---

[7] It seems that this practitioners' literature on gunnery (*Büchsenmeisterbücher*) was a particular German phenomenon that had no counterpart either in Italy or France at the time; see Leng (2004) 87ff. In fifteenth-century Germany, one encounters also technological literature written by practitioners for fellow practitioners in the field of architecture; for these *Werkmeisterbücher*, see sect. 4 of the chapter on architecture in the present book. For sixteenth-century gunners' manuscripts in England, see Walton (2000) appendices II and III.

[8] For the profession of the *Büchsenmeister* and particularly his position in and alongside the guild system, see, for instance, Leng (1996).

[9] For the eldest *Büchsenmeister* manuscripts, see Jähns (1889) book III § 57; Leng (2010) No. 39.1.

[10] For a translation of the *Feuerwerksbuch* into modern German, see http://www.feuerwerkbuch.homepage.t-online.de/downloads/FWB_neu.pdf. – For the preparation of gunpowder and its ingredients as well as the introduction of corned powder in the course of the fifteenth century, see Partington (1960) Chap. 4 and B.S. Hall (1997) Chaps. 2 and 3. For fifteenth-century manuscripts on gunpowder recipes, see Jähns (1889) book III § 63.

The *Feuerwerksbuch* of 1420 was widely distributed in the fifteenth and sixteenth centuries: More than 40 copies are known; it was translated into Italian and French; it was wholly or partly incorporated into other manuscripts on gunnery, e.g. Martin Merz's *Kunst aus Büchsen zu schießen* (c. 1475); and it was reprinted as late as in 1529, that is, 6 years before it was finally replaced as a standard text on gunnery by Franz Helm's *Buch von den probierten Künsten* (1535).[11]

Like other *Feuerwerksbücher*, Helm's book[12] largely consists of older texts and illustrations, among others of the *Feuerwerksbuch* von 1420 with its twelve canonical gunners' questions. These materials were, however, reworked and supplemented according to the technical developments since the fifteenth century. Notwithstanding such reworkings, the core content, as well as the basically practical character of Helm's and other contemporary gunners' manuscripts, remained unchanged well into the seventeenth century.

Despite an unbroken practice of producing and using gunners' manuscripts,[13] the sixteenth century brought a manifold printed literature on gunnery and ordnance that slowly changed the character of the technological literature in this field of practice. This holds even for the sequels to the gunners' manuscripts, namely the printed gunners' manuals of the time, as Steven Walton observed with regard to English gunner's manuscripts and printed manuals of the sixteenth and early seventeenth centuries:

> *Gunners' manual subject matter is quite uniform, even if the details vary widely from one source to the next. Gunners noted powder charges for various classes of cannon, the sizes of these cannon and their shot, recipes for mixing various types of gunpowder, and construction methods for different sorts of fireworks, a term which included incendiary weapons as well as aerial and ground-effect crowd-pleasers like those still used at public spectacles. Less surprisingly, they are also very interested in the sizes and construction methods of ladles for loading the cannon, and for good reason. Compared to the printed gunnery books, the manuscript tradition emphasizes the practical over the big picture. Much of the information is the same (charge weights, and ranges), but much that is minimized in books is emphasized in manuscript (ladles and, curiously, fireworks).*[14]

[11] For the known manuscripts of the *Feuerwerksbuch* and its continuance in several later manuscripts on gunnery, see Leng (2010) No. 39.2. For its translation into French, see Jähns (1889) book III § 60. The first print appeared in Augsburg from Heinrich Stainer in 1529. – Martin Merz: *Kunst aus Büchsen zu schießen* (Munich BSB Cgm 599).– Like the *Feuerwerksbuch* of 1420 one hundred years earlier, Franz Helm's *Buch von den probierten Künsten* initially existed only as manuscript (Heidelberg Cod. Pal. germ. 128, of which more than 70 copies are known) before it was printed with the title *Armamentarium principale oder Kriegsmunition und Artillerie-Buch* (Frankfurt/M.: Johann Ammon, 1625).

[12] For the following, see Leng (2010) No. 39.9.

[13] In the second half of the sixteenth century, when Helm's *Buch von den probierten Künsten* became a kind of reference text like the *Feuerwerksbuch* of 1420 a century earlier, still other gunners' manuscripts were composed; see Jähns (1889) book IV §§ 55 f.

[14] Walton (2000) 133. Walton transcribed Richard Wright's *Notes on Gunnery* of 1563 as well as the anonymous manuscript *The Secret of Gunmen* (Ashmole 343 temp James I), and edited these texts as appendices to his book.

Yet before we investigate this new technological literature, we will take a short look at an aspect of gunnery that had been excluded from gunners' manuscripts in the fifteenth century and was taken up again by writings on gunnery in the sixteenth century, namely the topic of manufacturing guns.

4.3 Excursus: Manufacturing Guns

The reason why gunners' manuscripts of the fifteenth century did not generally address the manufacturing of guns was the separation of the production and operation of guns at the end of the fourteenth century. Both activities had originally been carried out by gunsmiths when all kinds of guns were fabricated by wrought-iron work. With the application of metal casting in gun production, foundrymen took over the task of manufacturing some kinds of guns, namely the heavy pieces that were increasingly demanded in siege warfare. These highly specialized craftsmen did not usually take an active part in military campaigns, whereas the operators of these pieces, the gunners, no longer needed to be familiar with their fabrication.[15]

However, we still find some foundrymen among the authors of gunners' manuscripts. Martin Merz, for instance, had been a master foundryman before he advanced to become the commanding master gunner of the Palatinate under Elector Friedrich I. Merz, however, like other authors of *Büchsenmeisterbücher*, did not address the topic of gun fabrication in his gunners' manuscript *Kunst aus Büchsen zu schießen* (ca. 1475). One of the rare descriptions of the techniques of casting bells, guns, and sculptures before Biringuccio was composed by the foundryman Christoph Sesselschreiber (fl. 1520), who also became a master gunner.[16]

The Sienese engineer, mining official and head of the mint, Vannoccio Biringuccio (1480–1537), who went on to serve at the Papal court as an architect, head of the foundry and master gunner, composed a comprehensive treatise on metallurgy and metalworking which was published posthumously with the title *De la Pirotechnia* in 1540. This work became the standard reference on this topic for the next 200 years.[17]

---

[15] The art of metal casting had been an advanced technique in the ancient world and did not fall in oblivion during the Middle Ages, as Theophilus Presbyter's famous manuscript of ca. 1100 testifies (Wolfenbüttel HAB Cod. Guelf. Gud. Lat. 69 2°). The craft of founders had developed during the Middle Ages mainly in connection with bell founding. For the mutual adaption of improved gun powder mixtures and the size of guns in the fifteenth century, see B.S. Hall (1997) 87 f., who mentioned a list in which Leonardo da Vinci assigned different sorts of gunpowder to types of guns according to their dimensions (Leonardo (1974) Codex Madrid II fol. 98).

[16] For Merz' gunners' manuscript (Munich BSB cgm 599), see Jähns (1889) book III § 61 and Leng (2010) No. 39.6. Gun casting is described in part 2 of Christoph Sesselschreiber's illuminated manuscript on bell and gun casting – *Von Glocken- und Stuckgießerei* – from around 1520 (Munich BSB cgm 973).

[17] Vannoccio Biringuccio: *De la Pirotechnia*. Venice: Venturino Rufinelli, 1540. H.S. Mudd edited a modern English translation (and introduction) by Cyril Stanley Smith and Martha Teach Gnudi

Biringuccio's *Pirotechnia* includes descriptions of manufacturing glass, distilling acids and making fireworks alongside the whole range of sixteenth-century metallurgical and metal-working processes, such as smelting ores, assaying, alloying, and casting. The materials and processes related to bronze casting are a central topic of the work. The alloying of bronze is discussed in book V Chap. 10 and book VII Chaps. 5 and 6, while book VI provides an elaborate description of the preparation of molds for casting cannons and bells. Biringuccio also addresses topics like gun carriages (book VII Chap. 8), iron casting of cannonballs (book VII Chap. 9), loading and operating guns (book X Chap. 3), and gunpowder making, including the preparation of saltpeter (book X Chaps. 1 and 2).

In view of the attempts at establishing a theory of external ballistics which began in the sixteenth century with Tartaglia (see below), it seems appropriate to add some short technical remarks on the cannons as fabricated in the days of Biringuccio. Notwithstanding a broad variety of types of cannons still in use, in the mid-sixteenth century, "[…] the 'classic' form of large gun – the single-piece, cast muzzle-loading cannon – slowly came to predominate in every size category and caliber, every service (whether at sea or on land) […]" (B.S. Hall). This "classic" cannon had, of course, a smoothbore barrel (riffled ones first appeared in the nineteenth century), which meant that the shot's spin was unpredictable. For muzzle-loading guns, round shots (spherical cannonballs) were preferred, that is, shots with considerable aerodynamic drag. Furthermore, the developing predominance of this "classic" form of large gun should not be mistaken for a standardization of cannons and cannon balls; this means the performance of each single piece, whatever its type, could not be presupposed or anticipated, but had to be learned by operating it.[18]

4.4 Sixteenth-Century Gunners' Manuals and Treatises on Gunnery

It is well known that the rise of printed literature in the Renaissance was not just a technical event concerning the reproduction and distribution of texts (and images); rather, getting printed profoundly affected the parameters of texts – their readership, the circle of their authors, their commercial status, and thereby (more or less) noticeably their contents. The rise of printed literature on gunnery in the sixteenth century was no exception in this respect.

As regards the readership of printed books on gunnery, it is evident that these books were not just intended for gunners, as had been the case with the gunners' manuscripts of the fifteenth century. (As mentioned above, such manuscripts were

in 1943: Biringuccio (1959). In the sixteenth century, the work was frequently reprinted in Italy and translated into Latin (1540 and 1559), English (1555 and 1560), and French (1556 and 1572), see ibid., Appendix C; excerpts from this book can be found in several manuscripts and books on gunnery compiled in the sixteenth century; see ibid., Introduction, xxff. For the specific metallurgical contents of Biringuccio's treatise, see also sect. 3 of the chapter on chemistry in the present book.

[18] B.S. Hall (1997) 87 and 134ff.

produced – copied, compiled or written – and consumed alongside printed books well into the beginning of the seventeenth century.) In fact, printed books on gunnery usually targeted a broader audience – civilians as well as soldiers and military commanders.

This holds also for printed gunners' manuals, and not only with respect to newly composed ones like the *Büchsenmeisterey* (1591) by Franz Joachim Brechtel (1554–1593), but also for print versions of existing gunners' manuscripts like the *Feuerwerksbuch* of 1420 (printed in 1529) or of Franz Helm's *Buch von den probierten Künsten* (printed in 1625). Brechtel's entirely practical manual is a work of a practical mathematician (*Rechenmeister*) which specially emphasizes dimensions and measurements of barrels, shots etc., a topic that was certainly of interest not only for practicing gunners.[19]

A broader interest in the topics of traditionally gunners' manuscripts is also manifested by the practice of incorporating or appending compilations or adding excerpts of these manuscripts in printed books on warfare with a wide thematic scope, that is, to literature that generally focused on the "big picture," as Walton put it, rather than on the practical details that gunners needed to know.

Leonhard Fronsperger (c. 1520–1575), for example, incorporated his *Vonn Geschütz und Fewrwerck* (1557), a practical book on gunnery in which rockets and firework play the dominating role, into his three-volume *Kriegssbuch* of 1573 in which artillery issues form just a small part of a comprehensive treatise on all aspects of warfare - administrative, logistic, strategic, and technical.[20]

Walther Ryff's (1500–1548) *Buch der Geometrischen Büxenmeisterey* (1547) is a translation of book I and II of Tartaglia's *Nova Scientia* with a book on the usual topics of gunners' manuscripts appended to it. One comes across similar cases in *Arte of Warre* (1562) by Peter Whitehorne (fl. 1549–1563) and *Arte of Shooting* (1588) by Cyprian Lucar (1544–1611). Whitehorne's book is a translation of Macchiavelli's *Arte della Guerra* with an appended text – *Waies for the orderyng of Souldiers in Battelray* – addressing practical issues of siege warfare. Lucar's book is a translation of books I-III of Tartaglia's *Quesiti e inventioni diverse* with the addition of a text on the usual topics of gunners' manuscripts.[21]

---

[19] Franz Joachim Brechtel: *Büchsenmeisterey. Das ist: Kurtze doch eigentliche erklerung deren ding, so ein Büchsenmeister fürnehmlich zu wissen von nöten*. Nürnberg: Katharina Gerlachin, 1591. For Brechtel's book, as well as for printed books on gunnery of the second half of the sixteenth century, see Jähns (1889) book IV § 58.

[20] Leonhard Fronsperger: *Vonn Geschütz unnd Fewrwerck, wie dasselb zuwerffen und schiessen, Auch von gründlicher zuberaitung allerley gezeugs und rechtem gebrauch der Fewrwerck ...* Frankfurt/M.: Zephelius; idem: *Kriegssbuch*, 3 vol. Frankfurt/M.: Feyerabendt, 1573. For Fronsperger's books, see Jähns (1889) book IV §§ 47 and 52.

[21] Walther Ryff: *Das ander Buch der Geometrischen Büxenmeisterey*. Nürnberg: Johan Petreius, 1547; Peter Whitehorn: *The arte of warre, written first in Italia[n] by Nicholas Machiauell, and set forthe in Englishe by Peter Whitehorne, studient at Graies Inne: with an addicio[n] of other like marcialle feates and experimentes, and in a table in the ende of the booke maie appere.* S.l.: Niclas Inglande, 1562; Cyprian Lucar: *Three Bookes of Colloquies concerning the Arte of Shooting in great and small peeces of Artillerie*. London: Thomas Dawson, 1588. For Whitehorn and Lucar, see Walton (2000) 74ff. and 92ff.

Printed books on gunnery targeted not only a broader readership than gunners' manuscripts. They were also rarely written by gunners, who soon constituted a minority among the authors of books on gunnery. Among the authors of sixteenth-century books on gunnery we find military architects and engineers, e.g. Joseph Boillot (1546–1605), Girolamo Cataneo (1540–1584), and Luis Collado (fl. 1586), practical mathematicians, e.g. William Bourne (1535–1582) and Brechtel, astronomers like Daniel Santbech (fl. 1560), and even physicians like Walther Ryff. Mercenaries or military officials like Leonard Fronsperger were rather exceptions among these authors.[22]

Looking back to the subjects and genres of the sixteenth-century literature on gunnery, some deserve particular attention because they were to become of great significance in the future. Boillot's treatise, for instance, dealt with and assessed guns, gun carriages and mounting devices, as well as mechanisms of gun elevation based on the state of the art of mechanical engineering. French military commanders like the Grand-maître de l'*artillerie* de France, Jean d'Estrées (1486–1571), pushed in official papers for a standardization of calibers, a pressing topic in the time to come. Textbooks signaled that institutionalized teaching of the art of gunnery was nascent at the end of the sixteenth century: the *Instruttione de' bombardieri*, a textbook for a Venetian artillery school, appeared in 1592. And not a few approaches to a more theoretical treatment of gunnery were printed: for example, *An Arithmeticall Militarie Treatise, named Stratioticos* (1579), by Leonard Digges (c. 1515 – c. 1559) and his son Thomas (1546–1595), outlined the frame of a theoretical treatment of gunnery.[23]

> In[24] the section of *Stratioticos* titled *Certeine Questions in the Arte of Artillerie, by Mathematical Science joyned with Experience, to be debated and discussed*, Thomas Digges identified four "prime, substantial or effectuall causes" to which gunnery can be reduced: "powder, piece, bullet, and random" (i.e. the elevation of the piece) and added a number of "secondary or accidental causes", such as winds, wadding of the charge, fit of the shot to the bore, and the like. Several questions were attached to each of the "prime causes" that were, however, not answered but raised "to be debated and discussed;" e.g. as regards

[22] Joseph Boillot: *Modelles d'artifices de feu et divers instrumens de guerre utiles et nécessaires à tous ceux qui font profession des armes, avec les moyens de s'en prévaloir pour assiéger, battre, surprendre et deffendre toutes places*; Chaumont en Bassig: Q. Mareschal, 1598; – Girolamo Cataneo: *Opera nuova di fortificare, offendere et difendere*. Brescia: Giovanni Battista Bozola, 1564; – Luis Collado: *Pratica manuale di artigleria*. Venetia: Pietro Dusinelli, 1586; William Bourne: *The Arte of Shooting in Great Ordnaunce*. London: Thomas Woodcocke, 1587; Daniel Santbech: *Problematum astronomicorum et geometricorum*. Basel H. Petri & P. Pernam, 1561.

[23] Jean d'Estrées' official *Memoire de l'artillerie* and the *Traicté de l'artillerie* by Abra de Raconis with a similar thrust were not printed; d'Estrées was also co-author of a *Discours des villes, chasteaux, et fortresses batues, assaillies & prises, par la force de l'artillerie* (1563); see Jähns (1889) book IV § 60. – Eugenio Gentilini da Este: *Instruttione de' bombardieri, dove si contiene la Esamina usata dallo strenuo Zaccaria Schiavina*. Venetia: Francesco de' Franceschi, 1592. – Leonard Digges: *An Arithmeticall Militare Treatise, named Stratioticos ... Together with the modern militare discipline ....* London: Henrie Bynneman, 1579.

[24] For the following, see Walton (2000) 100ff.

"powder," what is the relation between distance and charge weights – linear, quadratic, or cubic (or no relation)?

Finally, it is evident that, in the course of the sixteenth century, more and more publications tried to treat aspects of gunnery in a mathematical way, not only determining distances, measures of guns and shots or gunpowder charges, but also, following the example of Tartaglia, the maximum range of guns and other issues of external ballistics.

However, before we embark on the field of external ballistics, we should ask – in view of the interplay between the technological and scientific literature of the early modern period – whether the literature on gunnery discussed so far profited from or was profitable for the scientific literature. Whereas it is hard to see that the literature on gunnery before Tartaglia took advantage of any scientific theories (apart from mathematics), there are two technological issues of gunnery that were taken up by the developing theory of chemistry. The first was the metallurgical knowledge gained and employed in the context of gun casting, and the second was the chemical knowledge gained and employed in the context of gunpowder production, particularly as regards the preparation of saltpeter, sulfur, and charcoal. (For the relationship of technological and scientific literature in the field of chemistry, see the chapter on chemistry in the present book.)

4.5 Tartaglia

In 1537, the practical mathematician and engineer Niccolo Tartaglia (1499–1557) published a treatise titled *Nova Scientia*. The inception of the *science* of external ballistics is usually dated to this work, which marks a turning point in the development of literature on gunnery. More precisely, as we will see, this was a turning point that gave rise to two different kinds of literature on gunnery. A closer look at this treatise may be appropriate.[25]

As regards the genre of the *Nova Scientia*, it can be taken as a typical practical mathematicians' text, that is, a treatise which employs resources of learned mathematics to discuss and solve problems pertaining to practical fields such as surveying, navigating, and so on or, in this case, the field of operating guns. Tartaglia had planned to divide the treatise into five books, but only three were actually published. The third book corresponds particularly well to the expectations for treatises by practical mathematicians. It gives a very instructive account of the fabrication, function, and use of a contemporary mathematical instrument for determining distances, the *geometrical quadrat*– not to be confused with the *gunners' quadrant* also

[25] *Nova scientia inventa da Nicolo Tartalea*. Vinegia: Stephano da Sabio, 1537. Further editions appeared in 1550, 1551, and 1558. A modern edition (a facsimile of the 1558 edition, transcription, and translation into English) appeared in Valleriani (2013). For the following, see Arend (1998) and Büttner (2017).

described by Tartaglia in the *Dedicatory Letter* of the treatise. (For these instruments, see sect. 4 of the chapter on practical mathematics in the present volume).

The first two books, however, are rather remarkable for a treatise by a practical mathematician. Here, Tartaglia presented a Euclidian-style deduction (definitions, suppositions, propositions) of the behavior of thrown "equally heavy bodies" (*corpi egualmente gravi*). Yet the resources on which Tartaglia drew for this purpose were not mathematical ones but came from natural philosophy: the Aristotelian distinction between natural and forced (violent) motions as well as assumptions involved in the theory of impetus. Based on these traditional principles of dynamics he constructed trajectories of projectiles with a peculiar composite path – a straight path in the beginning representing a violent motion and a straight perpendicular path in the end representing a natural motion, both interconnected by a circular curved path in the middle. (See Fig. 4.2) These path characteristics followed from the traditional dynamics applied by Tartaglia, particularly from the principle that no projectile can travel with a forced (violent) motion and a natural motion at the same time.[26] Thus, each and every trajectory – whatever its size, depending on the charge, and whatever its inclination, depending on the elevation – will conform to these specified path characteristics; and because of the lawful relation between elevation and trajectory established with these suppositions, Tartaglia could not only claim that the maximum range can be obtained with an elevation of 45° but furthermore, that the ranges any gun under scrutiny can obtain at each angle of elevation can be determined by one single test shot.

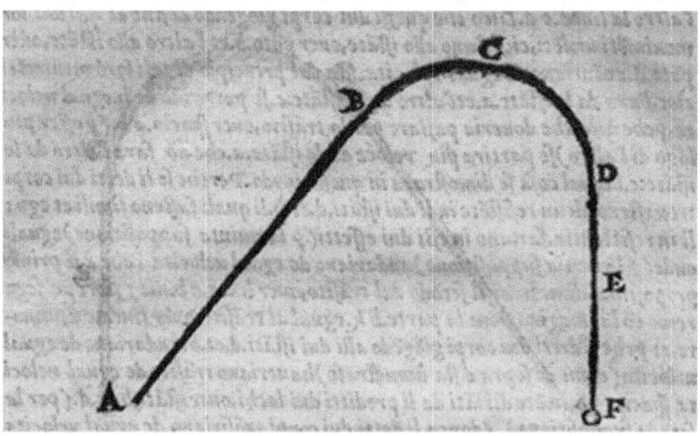

Fig. 4.2 Tartaglia's Trajectory (1558)
Niccolò Tartaglia: *Nova Scientia* I, fol. 7v

[26] *Nova Scientia*, book I, proposition 5: "*Niun corpo egualmente grave, puo andare per alcun spatio di tempo, ouer di loco, di moto natural, e uiolente insieme misto.*" For a discussion of controversial assessments by historians of science as to how strictly Tartaglia proposed and really adhered to this principle, see France (2014) Chap. 1.

It remains to add that the fourth book of the *Nova Scientia* Tartaglia had planned was thought to provide practical instructions on firing cannons and mortars over the whole range of elevations and furthermore, details on how to find the data for a *range table*; that is, tables that list the range of shots for every angle of elevation. Tartaglia incorporated some of these instructions into a later work, the *Quesiti e inventioni diverse.*[27]

One question regarding the relationship of Tartaglia's *Nova Scientia* and the early modern literature on gunnery is whether this treatise resulted from reflections on experiences and problems of contemporary gunners and would therefore help or even be indispensable for overcoming deficiencies in their practice. The relevance and feasibility of the issues in the focus of Tartaglia's ballistics are of particular interest for an assessment of the practical significance of the *Nova Scientia.* This ballistics' focus is on the range of shots, in particular on the question as to which elevation of the barrel would yield a maximum range. On closer inspection it seems rather dubious that this question was of great practical relevance at the time and it is also doubtful whether all of the elevations implied by Tartaglia could really be set up in practice.

In early modern siege warfare, cannons were employed for battering fortifications by direct fire ("at point-blank"). The same holds for the deployment of cannons on battle fields. The shorter the distance between gun and aim, the greater the shot's effect. In direct shooting, gunners found the effective angles of elevation by sighting and trying.[28] Indirect or "aimed" shooting across greater distances, and thus the question of the greatest range of a shot, was an almost negligible side issue of sixteenth- and seventeenth-century gunnery. Furthermore, the spectrum of angles to which gun barrels could be elevated was limited. Cannons could be elevated up to 45° degrees in theory only, but not in practice unless one dug a hole into which the tail of the gun carriage could dip.[29] (Fig. 4.3) This means that the assumed full spectrum of angles – from 0° to 45° for cannons, from 45° to 90° for mortars – was a geometrical fiction.

However,[30] although Tartaglia had no personal experience in this practical field, his approach to gunnery was not entirely amateurish or idiosyncratic. The range of shots were, as contemporary gunners' rules show, a concern of sixteenth-century gunners, albeit in an entirely different way, judging by these rules, which were

---

[27] *Quesiti e inventioni diverse de Nicolo Tartalea Brisciano*. Venetia: Ruffinelli, 1546.

[28] Büttner (2017) 123, cites a telling remark by William Bourne, who wrote: "First by their judgment they [sc. the gunners] doe give that so maney ynches advantage as they suppose will reach the marke, and then by the first lighting or falling of the shot, hee doth see whether it be shorte or gone over the marke, and if it be shorte, then at next shooting hee will give the peece more advantage and if it be over, then he will give the peece lesse advantage and so by divers times shooting off the peece at a marke, they will find how many ynches and partes will keep the length of the marke." (William Bourne: *The arte of shooting* (1587), see note 22 above.)

[29] See the quotation from the Dutch artillery commander Trolis Nielson Brinck's *Beschreyvinge van de artillereye* (1681) in Büttner (2017) 138.

[30] For the following, see Büttner (2017) 127ff.

Fig. 4.3 Elevation Limits Due to the Gun Carriage
Niccolò Tartaglia: *Nova Scientia* I, unnumbered

sometimes obvious and sometimes odd.[31] The promise that one specific parameter of shooting (the range of shots) could be controlled by one specific factor (the elevation) was certainly suggestive for practitioners. And the idea that geometrically constructed trajectories for each angle of elevation were the key to solving the range issue had been considered before Tartaglia, for example by Leonardo da Vinci,[32] and was pursued by writers on gunnery after him.

As regards the development potential of Tartaglia's *theoretical* approach to ballistics, ambivalent prospects can be stated.[33] On the one hand, by basing his geometrical construction of projectile trajectories on natural philosophical assumptions, Tartaglia not only opened up a new resource for treating ballistic issues, but gave birth to a new field of theoretical investigations – that of scientific external ballistics. On the other hand, by the implicit abstractions of his theoretical treatment of ballistics, he transformed the projectile and its trajectory into non-material entities. In his theory, as Jochen Büttner put it, shooting became "shooting with ink."

Tartaglia's reliance on dynamics in the tradition of Aristotle and the theory of impetus should not necessarily be seen only as a sign of backwardness. By

---

[31] Büttner, (2017) 122 f., quotes examples of such rules from the *Archiley-Kriegskunst* (1617), written by the head of the military college in Siegen, Johann Jacob von Wallhausen (ca. 1580–1627). For example, an obvious rule: "To shoot from a fortress / the higher the cannon is planted / the farther a shot can be made with it" and a strange one: "Shooting over grassland or dust / over water too / powder and elevation have to be increased / otherwise [the shot will be] much too short […]"

[32] Leonardo (1974) Codex Madrid I, f. 147r.

[33] For the following, see Arend (1998), Chap. 4, and Büttner (2017) 131ff.

employing these traditional dynamics theories for the investigation of new real-world phenomena, he initiated and provoked new conceptualizations of dynamics that led eventually to the modern dynamics of the seventeenth century (Galileo and Newton).

The raising of the projectile trajectory to the real subject of scientific external ballistics by Tartaglia's (and his followers') theoretical approach implied many abstractions –from the construction of guns, the material and shape of the shot, the explosive charge, the air resistance, and so on. The historical actors were more or less aware of these abstractions. But they were not aware that the shape, and even the path characteristics of projectile trajectories – that is, the path characteristics of Tartaglia's composite trajectory as well as, later, that of parabolic or hyperbolic trajectories – depend equally on the elevation of the gun and on the air resistance, which relies in turn on the muzzle velocity of the shot (its velocity when leaving the barrel). In other words, they were not aware of a fundamental flaw in their theoretical building that impaired its applicability to real ballistics.

Considering these strengths and weaknesses, promises and limits on both the practical and theoretical side of Tartaglia's approach, an ambivalent reception can be expected. Indeed, this ambivalence is present in the technological literature on gunnery as well as in the literature pertaining to the new scientific external ballistics – that is to say, ambivalent receptions in the two kinds of literature on ballistics that took different trajectories for the next century and a half.

4.6 External Ballistics from Tartaglia to Galileo

As regards the technological literature, one finds a wide spectrum of reactions to Tartaglia's theory, ranging from the one extreme of ignoring it altogether, to picking up certain of its statements and juxtaposing them with unchanged traditional issues, up to using it for special technical purposes.

External ballistics was not a topic of gunners' manuscripts or manuals before Tartaglia and there was no significant change in this respect in the sixteenth and seventeenth centuries. It is hardly surprising that issues relating to external ballistics and, thus, any reference to Tartaglia is missing from a work mentioned above, Franz Helm's *Buch von den probierten Künsten*, because it was composed at almost the same time as Tartaglia's *Nova Scientia*. However, Helm's book, which circulated in over 70 manuscript copies in the sixteenth century before it was finally printed in 1625, was one of the most influential gunners' manuals in continental Europe up to the mid-seventeenth century. Another telling instance of gunners' manuals that completely ignored external ballistics is *Ars Magnae Artilleriae* (1650) by Kazimierz Siemienowicz (c. 1600–c. 1651). It was translated into French, German, and English and remained the standard reference work on gunnery and the duties of master gunsmiths (*Büchsenmeister*) well into the eighteenth century. Books like these seem to confirm the minor importance of indirect or "aimed" shooting in this period.[34]

Among sixteenth-century books on gunnery that were compiled from different sources, not a few include parts or excerpts from Tartaglia's writings. As mentioned above, Walther Ryff incorporated a translation of books I and II of the *Nova Scientia* together with compilations from contemporary gunners' manuscripts in his *Geometrischen Büxenmeisterey* (1547). More frequently, one finds resonances, excerpts or even translations of or from Tartaglia's *Quesiti e inventioni diverse* in gunners' manuals. Cyprian Lucar's *Three Bookes of Colloquies concerning the Arte of Shooting* (1588), for instance, contains a translation of books I to III of the *Quesiti* and a large appendix compiled from various texts on gunnery by "diverse good authors in diverse languages." Often, however, it is not clear why the compilers of such manuals included materials relating to Tartaglia: was it to prove they were up to date or because of a genuine interest in issues of external ballistics? In the latter case, the practically oriented *Quesiti* were certainly more attractive for authors of gunners' manuals than the natural philosophical *Nova Scientia*.[35]

This observation is confirmed by two important gunners' manuals that treated issues of external ballistics, Luis Collado's (fl. 1580s), *Pratica manuale di artigleria* (1586), and Diego Ufano's (fl. 1610s), *Tratado dela artilleria y uso della platicado* (1613). Both authors elaborated Tartaglia's ideas mathematically for range tables as indicated in his *Quesiti*; but neither of them took up the natural philosophical underpinnings from Tartaglia's *Nova Scientia*.[36]

Manuals like that of Collado and Ufano show that external ballistics was not as glaringly absent from the sixteenth- and seventeenth-century technological literature on gunnery as the manuals by Helm or Siemienowicz may suggest, alongside the persisting marginality of indirect shooting in practice. Whatever their benefit for the practice of gunners, range tables, with or without geometrical constructions of projectile trajectories, increasingly became a topic in the technological literature on gunnery from the mid-sixteenth century on.

> These range tables must be seen against the background that guns were not yet subject to standardization in the early modern period; that is to say, the performance of each singular gun had to be established by testing in order to select the right table for it. (Related to this was the issue of the weight of fitting shots which had to be determined by means of caliber scales.[37])

---

[34] Kazimierz Siemienowicz: *Artis Magnae Artilleriae pars prima*. Amsterdam: Jansson, 1650.

[35] For the books by Ryff and Lucar, see note 21 above.

[36] Luis Collado: *Pratica manuale di artigleria*. Venetia: Pietro Dusinelli, 1586. Diego Ufano: *Tratado dela artilleria y uso della platicado*. Brussels: Momarte, 1613.

[37] Caliber scales were metal rules for measuring the diameter of a gun's barrel which indicated the corresponding weight of a stone, iron or lead shot. The first of such caliber scales was invented by the practical mathematician and instrument maker Georg Hartmann (1498–1564) in 1540; see Long (2012) 10.

> As[38] regards the methods of calculating range tables, arithmetical and geometrical ones both existed. By the *arithmetical* method, the table's range values are calculated by adding/subtracting fixed or regularly increasing/decreasing values with each subsequent angle of elevation. The tables by Ufano, Daniel Elrich (fl. 1670s), and Johann Siegmund Buchner (fl. 1680s) were calculated in this way, for example.[39] With the *geometrical* method, the values are trigonometrically calculated with reference to a geometrical representation of the two straight paths of a Tartaglia-type trajectory that form a right-angled triangle together with the horizontal straight line; the perpendicular path is taken as the sine of the angle of elevation and its cosine accordingly as the range. One of the first who applied this method was the Dutch astronomer Daniel Santbech (fl. 1560s) – (Fig. 4.4). Many others followed with variations, including Luis Collado, Diego D'Alaba y Viamont (1557-?), Cyprian Lucar, and David Rivault (1571–1616). [40]

It is important to emphasize that the external ballistics manifested by these range tables did not involve any attempts to model the projectile trajectory on the basis of natural philosophical concepts. None of the aforementioned authors can be seen as a follower of Tartaglia in this respect. This is obvious as regards those authors who followed arithmetic calculations of ranges. And as regards those followed the geometrical method, we should note that the trajectories constructed for the trigonometrical calculation of ranges resulted from mathematical rather than dynamic considerations, as is clearly visible in the case of Santbech (Fig. 4.4). Thus, treatises like Collado's or Ufano's that deal with issues of external ballistics are technological treatises rather than scientific ones and should not be regarded as treatises of the new *science* of external ballistics.[41] The center of this science was the derivation of projectile trajectories from principles of dynamics.

It took until 1638, almost exactly one hundred years after the publication of Tartaglia's *Nova Scientia* in 1537 before the science of external ballistics, that is, the efforts at conceptualizing the trajectory of thrown bodies on the basis of dynamical

[38] For the following, see Arend (1998) Chap. 5.

[39] It is of interest to notice that Elrich's as well as Buchner's tables (1676 and 1682, resp.) were appended or attached to editions of Siemienowicz's influential gunners' manual (e.g. the German edition of 1676) which did not, as mentioned above, deal with external ballistics. For the arithmetical method, see also François Blondel's (1618–1686) interpretation of Ufano's table (reproduced in Büttner (2017) 142). For such calculations, see also A.R. Hall (1959) 45ff.

[40] For Santbech's geometrical method, see, for example, Arend (1998) 79 and Stewart (2012) 163 who commented: "As physically unrealistic as a right-angled triangular trajectory may initially seem […] [this] practice seems less absurd when one realizes that straight-lined trajectories with abrupt acute angles were employed more for their calculational convenience than as a true description for the trajectory. […] So if the trajectory … could be approximated by simple straight-lined trajectories, its range could be found using nothing more than a table of sines." – Diego D'Alaba y Viamont: *El perfecto capitán instruido en la disciplina militar, y nueva ciencia de artillería* (Madrid: Por Luis Sanchez, 1590). For David Rivaut, see Arend (1998) 304ff.

[41] This also holds for some gunner's manuals that endorsed Tartaglia's claims more extensively, for instance the aforementioned *Stratioticos* by Leonard and Thomas Digges, or William Bourne's *The Arte of Shooting in Great Ordnaunce* (1587). Bourne, in his short Chap. 9, touched upon the projectile trajectory by just enumerating trajectories by six selected angles of elevation. For Bourne's familiarity with Tartaglia's *Quesiti*, see particularly Chaps. 6–9 of his *Arte of Shooting* and Walton (2000) 87.

Fig. 4.4 Santbech's 'Trajectory (1561)
Daniel Santbech: *Problematum astronomicorum et geometricorum,* p. 212

principles, reached a new stage with the publication of Galileo's *Discorsi e Dimostrazioni Matematiche intorno a due nuove scienze*. In the *Fourth Day* of this work, Galileo presented the theory that such bodies travel on parabolic trajectories. As is well known, Galileo's theory of parabolic projectile trajectories rested on three breakthroughs in dynamics that constituted cornerstones of modern dynamics: (1) the conceptualization of the law of free-fall, (2) a first version of a concept of inertial motion, and (3) the concept of mixed motion, that is of a simultaneous but independent action of two separate motions, an inertial and a gravitational one.

It lies outside the scope of the present work to trace the development of this new dynamics in the hundred years after Tartaglia's *Nova Scientia* – a development which was far from straightforward, comprising detours and deadlocks and proceeding through strange paths. This development remained largely invisible to contemporaries since almost no publication made known anything about it, with the curious consequence that even the two main protagonists of this development – Thomas Harriot (1560–1621) and Galileo Galilei – had almost no knowledge of, let alone any contact with, one another. Although both men not only studied ballistic issues but were at times actively engaged in such matters, it goes without saying that gunners' manuals and technological writings on artillery topics had little to offer they could exploit for their achievements in relation to the new dynamics.[42]

[42] While this holds true for most of the contents of these manuals, it has to be added that Harriot, as Matthias Schemmel has shown (Schemmel (2008) vol. I, Chaps. 2.2, 8.4 and 8.5), did make use of range tables given in gunners' manuals, e.g. in William Bourne's *The Arte of Schooting*: He com-

If it is dubious whether the external ballistics of the authors of range tables was of significance for gunners' practice, this seems even more doubtful in the case of Galileo's theory which studies the path of a moving body (represented by a geometrical point, albeit a gravitating one) in empty space. For Galileo's theory abstracted from the shape and spin of real projectiles, from inequalities of real guns, real battle grounds, real charges, from recoil, and so on – in other words, from the real conditions of shooting that gunners cannot abstract from. It is therefore not surprising that the improvement achieved by Galileo as regards a dynamical derivation of projectile trajectories did not mean an equivalent improvement to this theory's practical utility.

Evangelista Torricelli (1608–1647), Galileo's successor as court mathematician of the Medici, who made a major contribution to the propagation of Galileo's theory of ballistics through his treatise *De motu gravium naturaliter descendentium* (1644), was confronted with the discrepancy between theoretically derived and actually obtained ranges of shots by a practitioner, a gunner named Giovanni Battista Renieri. The ensuing correspondence between Toricelli and Renieri deserves attention because the former, though not really familiar with gunnery, identified some of the real circumstances of shooting from which the scientific external ballistics abstracted, as possible reasons for such discrepancies.[43]

Galileo's scientific theory of external ballistics, which became canonical for the ensuing decades, was deficient not only in its applicability to practical concerns of gunners. It was also seriously hampered by its abstractions with regard to dynamics, particularly the abstraction of the resistance of the air.

Galileo was not unaware of the resistance the air exerted on moving bodies. He studied the issue mainly in connection with the law of free fall and concluded that the law holds exactly only for bodies falling through a vacuum. As regards moving projectiles, however, he believed that the resistance of the air was negligible. This is, indeed, more or less true for shots with a very low velocity, for instance for mortar shots but not, however, for cannon shots. The neglect or underestimation of the resistance of the air impaired not only the practical value of Galileo's ballistics but also its main achievement, that is, the derivation of a parabolic trajectory.[44]

---

pared them with the ranges following from his dynamical theory and tried to explain the deviation of his own ranges from the allegedly measured ones of the tables with changes of the muzzle velocity due to the elevation of the gun. – For Harriot's dynamics, see also Walton (2000) Chap. 1, especially 17ff. and Appendix I, and France (2014) Chap. 3. For the development of Galileo's dynamics see, besides the classical studies of Stillman Drake, Renn et al. (2001) and Büttner (2019).

[43] See Segre (1983) and Büttner (2017) 159.

[44] As mentioned above, the path characteristics of trajectories depend, no less than on the elevation of the gun, on the air resistance which depends for its part on the muzzle velocity of the shot. The fact that Galileo's ballistics holds for slow mortar fire was probably the main reason why it became the canonic theory also for practitioners. Bernard Forest de Belidor (1697–1761), for instance, published range tables for mortars in his *Le bombardier français* (1731) which were calculated according to Galileo's theory.

In[45] the first half of the seventeenth century, the issue of the resistance of the air became a topic of interest independently of ballistics, namely as one aspect of the incipient aerostatics and aerodynamics of the time, and particularly in connection with the highly ideologically charged vacuum question. In other words, the issue of the resistance of the air was embedded in natural philosophical discourses rather than in those about technical questions even though almost all of the participants in these discourses were interested in such questions no less than in philosophical ones –Isaac Beekman (1588–1637), Marin Mersenne (1588–1648), René Descartes (1596–1650) and later Christian Huygens (1629–1695), to name just the most prominent ones. All attempts at calculating the effect of the medium air on the trajectory of projectiles made in the first two-thirds of the seventeenth century were, however, hardly satisfactory due to the limited mathematical means available before Leibniz and Newton invented the infinitesimal calculus. The mathematician, engineer, and soldier François Blondel (1618–1686), author of the then most esteemed French treatise on ballistics, *L'Art de Jetter les Bombes* (1683), and an adherent of Galileo's parabolic trajectory, doubted the possibility of exactly calculating the effect exerted by the air upon the projectile trajectory but attached little practical importance to this issue:

> *[…] l'on ne doit pas presumer que la résistance de l'air apporte de grands changements dans les mouvements des nos projections ordinaires.*[46]

4.7 Experimental Ballistics

The situation changed with the establishment of educational institutions for artillerists in several nations in the eighteenth century.[47] Some of these institutions became sites where ballistic problems, including the issue of the resistance of the air, were experimentally investigated. The experiments carried out mostly concerned technical questions such as reduced gunpowder charges and, as a result, lighter guns, their fabrication, particularly the problem of precise and standardized boring of gun barrels and the production and standardization of fitting shots. These experiments resulted in technical improvements that considerably enhanced the predictability of shots, narrowing the gap between theoretical and practical ballistics.[48]

---

[45] For the following, see A.R. Hall (1952) Chap. 5.

[46] François Blondel: *L'Art de Jetter les Bombes*. Paris: chez l'autheur et Nicolas Langlois, 1683, 345, quoted after A.R. Hall (1952) 127.

[47] France: Artillery schools at the end of the seventeenth century in Douai and Metz, artillery academy 1719 in La Fere; England: *Royal Naval Academy* Portsmouth 1722, *Royal Military Academy* in Woolwich 1741; Prussia: *Academie Militaire* (1765) which, however, neglected the artillery; Hapsburg: *Technische Militärakademie* 1717 which focused on military engineers rather than artillerists.

[48] Steele (1994) 354. Particular attention is merited by the invention of drilling machines for cast guns by the Suisse gun founder Jean de Maritz (1680–1743) – a vertical drilling machine in 1713 and a horizontal one in 1734 which gave perfectly straight bores exactly fitting the ball diameter.

There[49] is an English prelude to experimentation on ballistic issues in artillery academies that deserves to be mentioned. In 1674, Robert Anderson (fl. 1668–1696)[50] published *The Genuine Use and Effects of the Gunne*. In this gunners' manual, Anderson, like Galileo, regarded the resistance of air as negligible. This provoked a sharp reaction from James Gregory (1638–1675) who had published his own attempt at calculating the effect of the resistance of the air on projectiles only 2 years earlier in his *Tentamina quaedam Geometrica de Motu Penduli & Projectorum*.[51] The ensuing controversy was joined by excellent mathematicians like John Wallis (1616–1703) and Isaac Newton, and finally prompted the Royal Society to carry out some shooting experiments with the assistance of the Master of the Ordnance. (This seems to have been the first occasion that Newton became occupied with the issue of the resistance of air, an issue to which he would later dedicate book 2 of his *Philosophiae naturalis principia mathematica*.) Another such prelude is worth mentioning: In the 1660s, Lord William Brouncker (1620–1684) made experiments on the troublesome issue of the recoil of guns during a series of experiments conducted by the Gresham College group a precursor of the Royal Society.[52]

Groundbreaking experiments regarding the resistance of the air were first conducted in the 1740s by Benjamin Robins (1707–1751). He published the results in *New Principles of Gunnery* (1742), which included severe qualifications of Newton's and Huygens' derivation of the ratio of air resistance and projectile velocity as well as a complete refutation of Galileo's parabolic trajectory. These results initiated the so-called Ballistic Revolution of the eighteenth century. (Fig. 4.5).[53]

Key to Robins' achievements was the invention of two instruments that allowed the determination of the velocity of shots – their muzzle velocity as well as that at intervals during their flight. One instrument was the *ballistics pendulum* and the other the *whirling arm* (which is sometimes seen as a forerunner of modern-day wind tunnels). By means of these devices, he was able to deliver the numerical values lacking in all the earlier derivations of the projectile trajectory, which allowed calculation of the force of the air resistance and its impact on the trajectory on a new basis.

A translation of Robins' treatise into German, translated and extensively commented by Leonhard Euler (1707–1783), appeared as early as 1745.[54] Euler's

---

[49] For the following, see A.R. Hall (1952) 120ff.

[50] In the last three decades of the seventeenth century, Anderson, a London silk-weaver and mathematician, carried out a huge series of experiments by himself on shooting with cannons.

[51] The *Tentamina* is an appendix to Gregory's *The great and new art of weighing vanity.*

[52] "Experiments of the Recoiling of Guns by the Lord Brouncker," in *Sprat* (1667) 233–239.

[53] Benjamin Robins: *New Principles of Gunnery: Containing the Determination of the Force of Gun-Powder, and an Investigation of the Difference in the Resisting Power of the Air to Swift and Slow Motion*. London: J. Nourse, 1742. As this title indicates, Robins also made important contributions to the development of internal ballistics. For Benjamin Robins and the 'Ballistic Revolution' see, for instance, Steele (1994) and Barnett (2009).

[54] Leonhard Euler: *Neue Grundsätze der Artillerie*. Berlin: A. Haude, 1745.

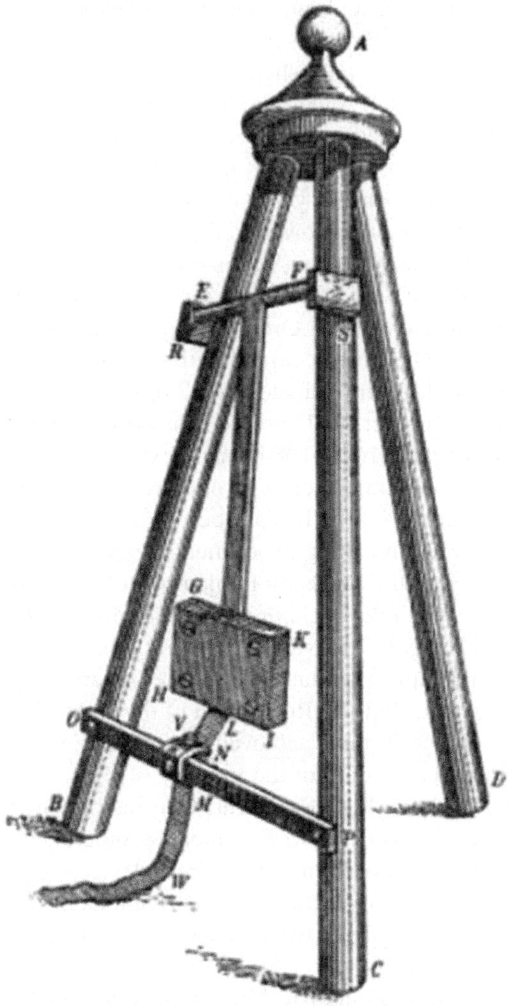

Fig. 4.5 Benjamin Robins' Ballistic Pendulum Robins and Euler (1745) Table II

comments were partly critical but generally praised Robins' achievements. Regarding the *ballistics pendulum*, Euler appreciated it as "one of the most ingenious and useful discoveries in artillery," adding that all artillery experimentation before Robins "was not only uncertain but erroneous."[55] As regards Euler's own exploitation of this new basis of scientific external ballistics, it took another 8 years before he published his *Recherches sur la véritable courbe que décrivent les corps jettés dans l'air ou dans un autre fluide quelvonque*. According to Janet Barnett, this was a "complete analysis of equations for ballistic motion in a resisting medium."[56]

[55] Quotation after Johnson (1992) 672.

[56] Euler *Opera omnia*, II/14 (Berlin 1922) 413–447; Barnett (2009) 99.

4.8 Conclusion

To sum up, it seems that the technological literature on gunnery and the scientific literature on ballistics developed largely independently of each other in the early modern period. True, the wealth of experiences with firearms constituted the background against which the development of early modern external ballistics should be seen. But the pioneers of a scientific external ballistics – Tartaglia, Harriot, Galileo – found almost nothing in the contemporary literature on gunnery that they could use for their attempts at deriving the path characteristics of projectile trajectories from dynamic principles. And conversely, the most influential gunners' manuals from the sixteenth and seventeenth centuries, like those of Franz Helm or Kazimierz Siemienowicz, could obviously afford largely to ignore issues of external ballistics, since indirect shooting remained a practically irrelevant topic of gunnery well into the eighteenth century. The early modern literature on gunnery and the literature on "shooting with ink" traveled on separate trajectories.

In the course of the seventeenth century, however, there was a point of interaction between these two bodies of literature in relation to attempts at calculating range tables based on dynamically derived trajectories of projectiles (not, as in the sixteenth century, based on mathematically convenient diagrams of trajectories). The contact between the two literatures was initially unilateral. The military engineer Bernard Forest de Bélidor made use of Galileo's theory of parabolic projectile trajectories, although it held true only for very slow shots, as a resource for calculating range tables, whereas natural philosophers like Galileo, Huygens, and Newton found no comparably utilizable resource in contemporary range tables or gunners' manuals. These natural philosophers mainly lacked empirical data about shooting and were unable to find it in the technological literature. This unsatisfactory situation gradually changed with technical improvements to the guns (comparable performances etc.) and with the rise of experimental shooting. It was on this basis, finally, that Benjamin Robins found ways to deliver what the science of external ballistics needed for accomplishing the so-called Ballistic Revolution of the eighteenth century — the event that gave birth to a science of external ballistics that was both a technological and a natural science.

References

[Anonymus]. 1757. *Reglement für das Kaiserlich-Königliche gesamte Feld-Artilleriecorps*. Wien: Johann Thomas Trattner.

Arend, Gerhard. 1998. *Die Mechanik des Niccolò Tartaglia im Kontext der zeitgenössischen Erkenntnis- und Wissenschaftstheorie*. München: Institut für Geschichte der Naturwissenschaften.

Barnett, Janet Heine. 2009. Mathematics goes Ballistic: Benjamin Robins, Leonhard Euler, and the Mathematical Education of Military Engineers. *BSHM Bulletin: Journal of the British Society for the History of Mathematics* 24 (2): 92–104.

Biringuccio, Vannoccio. 1959. *The Pirotechnia of Vannoccio Biringuccio*. New York: Basic Books.

Büttner, Jochen. 2017. Shooting with Ink. In *The Structures of Practical Knowledge*, ed. Matteo Valleriani, 115–166. Cham: Springer.
———. 2019. *Swinging and Rolling: Unveiling Galileo's Unorthodox Path from a Challenging Problem to a New Science*. Dordrecht: Springer.
Damerow, Peter, and Wolfgang Lefèvre, eds. 1985. *George Adams: Geometrische und graphische Versuche. Nach der deutschen Ausgabe von 1795*. Darmstadt: Wissenschaftliche Buchgesellschaft.
Giorgio Martini, Francesco di. 1967. *Trattati di architettura, ingegneria e arte militaria (ca. 1485)*, ed. Corrado Maltese and Livia Maltese Degrassi. Milano: Il Polifilio.
France, Catherine Ann. 2014. *Gunnery and the Struggle for the New Science (1537–1687)*. PhD, School of Philosophy, Religion and History of Science, Univeristy of Leeds.
Hall, Alfred Rupert. 1952. *Ballistics in the Seventeenth Century: A Study in the Relations of Science and War with Reference Principally to England*. Cambridge: Cambridge University Press.
———. 1959. The Scholar and the Craftsman in the Scientific Revolution. In *Critical Problems in the History of Science*, ed. Marshal Clagett, 3–23. Madison: University of Wisconsin Press.
Hall, Bert S. 1979. *The Technical Illustrations of the So-Called "Anonymous of the Hussite Wars"*. Wiesbaden: Reichert.
———. 1997. *Weapons and Warfare in Renaissance Europe*. Baltimore: Johns Hopkins University Press.
Jähns, Max. 1889. *Geschichte der Kriegswissenschaften vornehmlich in Deutschland: 1. Abt. (Altertum, Mittelalter, XV. und XVI. Jahrhundert)*. München und Leipzig: R. Oldenbourg.
Johnson, William. 1992. Encounters between Robins, Euler and the Bernoullis: Artillery and Related Subjects. *International Journal of Mechanical Sciences* 34: 651–679.
Leonardo da Vinci. 1974. *Codex Madrid I*, ed. L. Reti et al. Frankfurt am Main: Fischer.
Leng, Rainer. 1996. Getruwelich dienen mit Buchsenwerk. Ein neuer Beruf im späten Mittelalter: Die Büchsenmeister. In *Strukturen der Gesellschaft im Mittelalter*, ed. Dieter Rödel and Joachim Schneider, 302–321. Wiesbaden: Reichert.
———. 2002. *Ars belli: Deutsche taktische und kriegstechnische Bilderhandschriften und Traktate im 15. und 16. Jahrhundert*. 2 vols. Wiesbaden: Reichert.
———. 2004. Social Character, Pictorial Style, and the Grammar of Technical Illustration in Craftsmen's Manuscripts in the Late Middle Ages. In *Picturing Machines: 1400–1700*, ed. Wolfgang Lefèvre, 85–111. Cambridge, MA: The MIT Press.
———. 2008. Zum Verhältnis von Kunst und Krieg in den illustrierten Kriegslehren des 15. und 16. Jahrhunderts. In *"Mars und die Musen": das Wechselspiel von Militär, Krieg und Kunst in der Frühen Neuzeit*, ed. Jutta Nowosadtko and Matthias Rogg, 33–59. Münster: LIT Verlag.
———. 2010. Feuerwerks- und Kriegsbücher (Nr. 39). In *Katalog deutschsprachiger illustrierter Handschriften des Mittelalters (KdiH)*, ed. Ulrike Bodemann, Peter Schmidt, and Christine Stöllinger-Löser. München: C.H. Beck.
Long, Pamela O. 2012. Trading Zones: Arenas of Exchange during the Late-Medieval/Early Modern Transition to the New Empirical Sciences. *History of Technology* 31: 5–25.
Marsden, Eric W. 1969. *Greek and Roman artillery – Historical development*. Oxford: Clarendon.
———. 1971. *Greek and Roman artillery – Technical treatises*. Oxford: Clarendon.
Partington, J.R. 1960. *A History of Greek Fire and Gunpowder*. Cambridge: Heffer.
Renn, Jürgen, Peter Damerow, and Simone Rieger. 2001. Hunting the White Elephant: When and How did Galileo Discover the Law of Fall. In *Galileo in Context*, ed. Jürgen Renn, 29–149. Cambridge: Cambridge University Press.
Robins, Benjamin, and Leonhard Euler. 1745. *Neue Grundsätze der Artillerie*. Berlin: Haude.
Santbech, Daniel. 1561. *Problematum astronomicorum et geometricorum sectiones septem*. Basel: H. Petri & P. Pernam.
Schemmel, Matthias. 2008. *The English Galileo: Thomas Harriot's work on motion as an example of preclassical mechanics*, Boston Studies in the Philosophy and History of Science. Vol. 1. Dordrecht: Springer.
Segre, Michel. 1983. Torricelli's Correspondence on Ballistics. *Annals of Science* 40 (5): 489–499.

Sprat, Thomas. 1667. *The History of the Royal Society, for the Improving of Natural Knowledge*. London: J. Martyn.

Steele, Brett D. 1994. Muskets and Pendulums: Benjamin Robins, Leonhard Euler, and the Ballistic Revolution. *Technology and Culture* 35 (2): 348–382.

Stewart, Séan M. 2012. On the Trajectories of Projectiles Depicted in Early Ballistic Woodcuts. *European Journal of Physics* 33: 149–166.

Tartaglia, Niccolò. 1558. *Nova scientia inventa da Nicolo Tartalea*. Vinegia: n.p.

Valleriani, Matteo. 2013. *Metallurgy, Ballistics and Epistemic Instruments*, Max Planck Research Library for the History and Development of Knowledge. Berlin: Edition Open Access.

Walton, Steven Ashton. 2000. *The Art of Gunnery in Renaissance England*. Ottawa: National Library of Canada.

Chapter 5
Mechanical Engineering

5.1 Pictorial Documents: Technological Literature?

The search for early modern technological literature on mechanical engineering reveals an odd situation: No such literature seems to exist before the end of the sixteenth century or even before the second half of the seventeenth century if one looks for descriptive, instructing, or analytic treatises. Instead one finds an abundance of pictorial documents – first, illuminated manuscripts and later, in the age of printing, the famous theaters of machines, documents in which texts accompany pictures (and not the other way around) and sometimes even documents that only contain pictures. Does it make sense to regard such documents as technological literature on mechanical engineering?

As in discourses on botanical or zoological specimens, no meaning can be transmitted by words alone in discourses on technical devices. Pictures (or diagrams) are indispensable as long as the audience is not already familiar with the subject at hand. Moreover, pictures may sometimes prove to be as insufficient representations as words and have to be complemented by physical representations in herbals, zoological collections, or in cabinets of models of machines or machine parts. Whether or to what extent late medieval illustrated manuscripts or early modern printed theaters of machines can or should be taken as technological literature depends not least on the audiences they were intended to inform or instruct.

The pictorial documents discussed here addressed potential customers as well as potential producers of the devices presented, and frequently both of these targeted recipients at the same time. Most of the illuminated manuscripts of the late Middle Ages presented devices employed in siege warfare (*poliocretics*) and addressed primarily princely or other military commanders, usually, though not always, to explain the use and function of the device at hand. They not only transmitted technological information of interest to today's historians of technology but were also significant for contemporaneous readers interested in technological matters. This holds even

W. Lefèvre, *Minerva meets Vulcan: Scientific and Technological Literature – 1450-1750*, Archimedes 60, https://doi.org/10.1007/978-3-030-73085-7_5

more for those documents that address the potential manufacturer of military or civil devices and also for drawings or descriptions drafted by technicians for the purpose of obtaining a sort of patent or privilege for an invention from a government body.[1] And even manuscripts or prints that seem to be nothing but a self-advertisement of the author may have functioned as technological information for contemporaries. It therefore seems well advised not to be over-hasty in removing these pictorial documents from the list of technological literature on mechanical engineering.

5.2 The Age of Illuminated Manuscripts

The two oldest of the known illuminated manuscripts on machines already show which information or instructions were transmitted by such pictorial documents. In 1335, the physician Guido da Vigevano (ca. 1280 – ca. 1350) composed an illuminated manuscript on military equipment to accompany the French king Philippe VI on his crusade. Since the proposed war machines and vehicles were not to be transported from France to the Holy Land but to be produced there, Guido provided careful instructions on how to manufacture the devices, invoking the opinion of the actual maker several times. This detail is of general interest since it draws attention to a specific character of late medieval and early modern pictorial manuscripts or prints on technical issues: instead of aspiring to completeness they focus on details that are not obvious to experienced makers and usually omit details that would be familiar to them (Fig. 5.1). At the beginning of the fifteenth century the physician Konrad Kyeser (1366–1405) composed an illuminated manuscript that depicted and described mainly military equipment. In contrast to Guido da Vigevano, Kyeser's focus was on the function and use of the devices rather than on their fabrication. His manuscript was primarily addressed to potential customers, which probably explains the astonishing spread of Kyeser's work.[2]

Despite differences in emphasis as regards the aspects described and explained, the illuminated manuscripts on machines were addressed equally to potential

---

[1] For patents or privileges before the establishment of the institution of patents in today's sense, see Popplow (1996) and (1998), 47ff. For the social context of early modern machine drawings, see Popplow (2004).

[2] Guido da Vigevano: *Texaurus Regis Francie* (Paris Bibliothéque Nationale de France MS lat 11,015; further copies: Codex Yale, Yale Center for British Art, Paul Mellon Collection, Mil. mss (4°), f. 8r; Torino Codex, Biblioteca Universitaria di Torino, ms. G.V. 9, pl. vii.) (DMD ID gdv). The physician Guido is also known for his *Anatomia*, a work on dissection that also contained illustrations. – Konrad Kyeser: *Bellifortis* (e.g. Göttingen, Niedersächsische Staats- und Universitätsbibliothek, 2° Cod. Ms. philos. 63) (DMD ID ky). More than 40 manuscript copies are known; for a list of manuscript copies of the *Bellifortis*, see Leng (2002), vol. II, 423–440. Both authors certainly drew on book IV of Publius Flavius Vegetius Renatus' *De re militari* (fourth century CE), a text that remained unforgotten during the Middle Ages. Kyeser became even more familiar with contemporary war machines as a participant in the crusades led by the Hungarian king Sigismund in the 1390s.

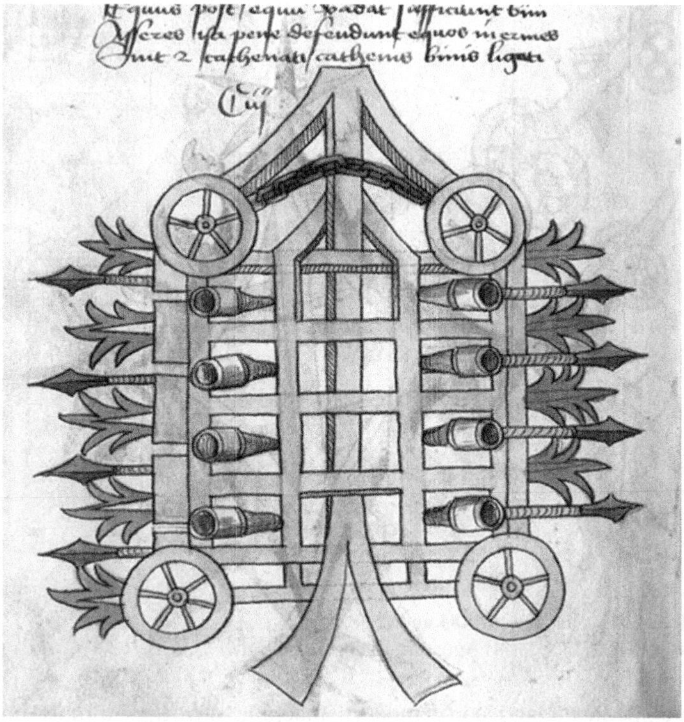

Fig. 5.1 Battle Wagon with Eight Guns (ca. 1430)
Konrad Kyeser: *Bellifortis* – BSB Clm 30150, f. 57v

makers and employers. Accordingly, these manuscripts had to develop a pictorial language that was intelligible to both classes of recipients. In this context, this language – which neither rendered exactly in a linear perspective, nor represented orthogonal plans – appears far from naïve and was rather a working compromise. This is corroborated by the fact that the theaters of machines composed and published as late as the turn of the sixteenth century essentially shared characteristics with the pictorial language achieved by late fifteenth century manuscripts; and this although rendering in perspective as well as constructing orthogonal plans had meanwhile been developed to high standards both in visual arts and architecture.[3] (Fig. 5.2)

[3] The "flat style" of rendering used by medieval authors like Guido da Vigevano or Conrad Kyeser suggests that they were probably familiar with ancient Byzantine or medieval Arabic technical drawings – for such drawings, see, for instance, Lefèvre (2002). Like modern orthogonal plans of machines, ancient and medieval drawings in flat style sometimes reach or overstep the limits of what is intelligible for laymen. Exact rendering in linear perspective, on the other hand, is unsatisfactory since it does not allow a single drawing to transmit all the information that makers need. The compromise found in the course of the fifteenth century is convincingly described by David McGee (2004). For the repertory of graphical "tricks" – exploded views, cutaway views, phantom

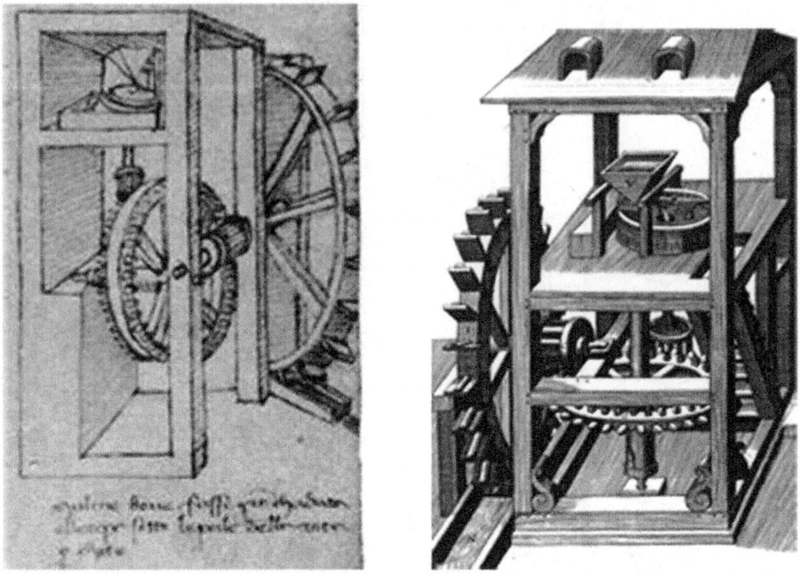

Fig. 5.2 Flour Mills – Depicted 1485 and 1588
Giorgio Martini (1967) I, p. 144; Agostino Ramelli (1588) f. 175r

The late medieval and early modern machine drawings almost immediately became collectibles and functioned as arsenals of war machines on paper at courts or sample collections of civil engineering contrivances maintained by urban and other administrative boards. The didactic value of such collections as illustrative materials for aspiring engineers was obvious. Some collections were built up solely for didactic purposes and can be regarded as forerunners of model collections of educational institutions like the mining academies of the eighteenth century.

Fifteenth-century illuminated manuscripts like that of Kyeser, Giovanni Fontana (1395–1455) or Roberto Valturio (1405–1475), although originally not composed for this purpose, were already used as such collections in that century. Illustrated manuscripts by gunsmiths like that by Johannes Formschneider (ca. 1420 – ca. 1470) or the illuminated inventories of arsenals by Ulrich Bessnitzer (ca. 1450–1521) and Philipp Mönch (fl. 1490) were composed as such arsenals on paper, as was the *Weimarer Ingenieurkunst- und Wunderbuch*, an anonymous work compiled around 1500. As late as the 1570s, Duke Julius of Braunschweig-Wolfenbüttel commissioned a collection of machine drawings intended, among other purposes, to provide the information and knowledge administrative officials needed for procuring the most advanced instruments for their area of responsibility. In the fifteenth century the *Spedale di Santa Maria della Scala* in Siena, originally a hospital and hostel for pilgrims, managed a good deal of the city state's infrastructure and housed a building workshop which held collections of plans, drawings, and models for civil

views, etc. – applied to overcome the limits of the given pictorial language, see Leng (2004) section 4 and Lefèvre (2003) 78ff.

engineering – a collection which may possibly be mirrored in the famous *Trattati* of Francesco di Giorgio Martini (1439–1501).[4]

5.3 The Age of Theaters of Machines

The features of fifteenth- and early sixteenth-century illuminated manuscripts and booklets on machines discussed so far are also characteristic of the technological literature on mechanical engineering published before the second half of the seventeenth century, particularly of the famous theaters of machines.

The series of these theaters opened with the *Theatrum Instrumentorum et Machinarum* (1578) by the practical mathematician Jacques Besson (1540–1573). In 1584 the practical mathematician, military engineer and soon-to-be *Premier ingénieur du roi* Errard de Bar-le Duc (1554–1610) published his instrument book, followed only four years later by Agostino Ramelli's (1531–1610) famous theater; Ramelli, too, was a practical mathematician and military engineer in the service of Henri d'Anjou, later King Henri III of France. This was followed in the first third of the seventeenth century by three theaters in Italy – composed by the architect and engineer Vittorio Zonca (1568–1603), the diplomat and (erstwhile) commander of a fortress, Fausto Veranzio (1551–1617), and the engineer and architect Giovanni Branca (1571–1645).[5]

---

[4]The remarkable spread of Kyeser's manuscript speaks for its use as a collectible. Giovanni Fontana: *Bellicorum instrumentorum liber*, 1430 (Bayerische Staatsbibliothek Munich Cod. Icon. 242) (DMD ID fong). Roberto Valturio: *De re militari libri xii*, 1466 (Torino, Archivo storico AMMA: Duca di Genova fol. (datato anno 1466) ff. 141) (DMD ID val), printed in Verona by Johannes Nicolai, 1472, several editions; copies of Valturio's manuscript were commissioned by his patron, Sigismondo Pandolfo Malatesta, duke of Rimini, to distribute to other rulers; Johannes Formschneider: [*Büchsenmeisterbuch*], ca. 1475 (Bayerische Staatsbibliothek Cgm 734) (DMD ID fs), several manuscript copies; Ulrich Bessnitzer: [*Zeughausinventar*] 1489 (Universitätsbibliothek Heidelberg cpg 130) (DMD ID bn); Philipp Mönch: *Dys büch der stryt vnd büchßen*, 1496 (Universitätsbibliothek Heidelberg cpg 126) (DMD ID mo). Anonymus: [*Weimarer Ingenieurkunst- und Wunderbuch]*, 1500 (Herzogin Anna Amelia-Bibliothek Weimar fol 328) (DMD ID wiw); Anonymus: [*Instrumentenbuch*] 1573 (part I: Niedersächsisches Staatsarchiv Wolfenbüttel, 2 Alt 5228, part II: Landeshauptarchiv Sachsen-Anhalt in Magdeburg Rep. Cop., Nr. 803b) (DMD ID jua, jub). – For the collections in Siena, see Elizabeth Merrill (forthcoming).

[5]Jacques Besson's *Theatrum Instrumentorum et Machinarum,* Lyon: Bartelemy Vincent, 1578; this was a posthumously edited print of an illuminated manuscript of 1569, supplemented with comments by François Béroalde de Verville (1556–1626); this edition was reprinted several times and translated into various vernaculars. – Errard de Bar-le-Duc: *Le Premier livre des instruments mathématiques et méchaniques*, Nancy: Jan-Janson, 1584. – Agostino Ramelli: *Le diverse et artificiose machine ... composte in lingua Italiana et Francese*, Paris, by the author, 1588. – Vittorio Zonca: *Novo Teatro Di Machine Et Edificii [...] Opera necesaria ad Architetti, et a quelli, / chi di tale studio si dilettano*. Padova: Pietro Bertelli, 1607. – Fausto Veranzio: *Machinae novae*. Venice: n.p., 1616. – Giovanni Branca: *Le machine: volume Nuovo [...] in lingua volgare et Latina*. Rome: Iacomo Mascardi, 1629. – In seventeenth-century Germany two theaters appeared which were largely compilations of materials from these French and Italian theaters: Henricus Zeising: *Theatrum machinarum* (1607–14) and Georg Andreas Böckler: *Theatrum machinarum novum* (1661).

Like the illuminated manuscripts about machines, these theaters are not treatises in which pictures help to understand a text but showcases of pictures (engravings) of machines or sites where machines were employed – pictures accompanied by (often very short) texts. They address, first and foremost, potential customers for the devices presented, e.g. princes and other military commanders. They try to induce these leaders to commission and deploy the machines (or to appoint the author). In a way these theaters are catalogues of devices the author presents as "new, useful, and ingenious."[6] Their emphasis lies on the use and function of a device rather than on how it can be fabricated or implemented.

The much-discussed question of whether the machines presented in these theaters were only imagined or actually real devices is pointless with respect to such catalogues of novelties. The theaters did not record machines made and employed at a specific work site and a specific place but presented proposals for or ideas about machines. If any of these proposals were to be realized, they first had to be adapted to the local circumstances of their employment; a one-to-one realization of a device as depicted is impossible.

There are some examples of early modern pictorial presentations of machinery that – in contrast to the theaters of machines – actually record existing and deployed machinery. One famous instance is Georgius Agricola's (1494–1555) presentation of the machinery used in the mining district of Joachimsthal in Saxony (now Jáchymov in the Czech Republic) in book VI of *De re metallica* (1556).[7] (Fig. 5.3) Equally famous are the drawings by Mariano di Jacopo [Taccola] (1381–1453) and Leonardo da Vinci (1452–1519) of the machinery built and employed by Filippo Brunelleschi (1377–1446) for the construction of the cupola of Florence cathedral.[8]

It is difficult to decide whether all – or some, or none – of the depicted devices could have been adapted to local circumstances and realized, apart from ubiquitous known und used devices like a winch or evidently unfeasible proposals such as a perpetuum mobile. The images in the theaters of machines are generally incomplete, that is, they do not depict all the details that would allow us to answer this question. As regards original and inventive proposals, this incompleteness could have been due to the author's interest in concealing details that might give him an

---

6 "Neu, nützlich und erfindungsreich" is the title of Popplow's work (1998), which includes a thoughtful discussion of the various interpretations of the theaters of machines (pp. 65–94).

7 The same cannot be claimed unambiguously for the *Schwazer Bergbuch* (1556) as regards its presentation of the machinery employed in the mining district of Schwaz (Tyrol); see Lefèvre (2010). For Agricola's presentation, see the chapter on mining science, section 3, in the present volume.

8 See DMD IDs tit10, tit11, LdVCA034, LdVCA035. Some of Brunelleschi's machines were also depicted in Francesco di Giorgio Martini's *Trattato I* (DMD IDs gm94a, gm95a, gm320), several copies of which circulated in the sixteenth century. These machines were also pictorially recorded by Bonaccorso Ghiberti (1451–1516) and Giuliano da Sangallo (1453–1516). The architect Antonio da Sangallo (DMD IDs sa1449va, sa1450vd, sa4078ra) or the engineer Heinrich Schickhardt (1558–1635) are also known to have recorded remarkable technical devices in private notebooks or private portfolios. It is, however, not unproblematic to regard these records as technological literature, although such materials were copied and circulated in the early modern age without being ever printed; see Merrill (2017).

A—Reservoir. B—Rack. C, D—Levers. E, F—Troughs under the water gates. G, H—Double rows of buckets. I—Axle. K—Larger drum. L—Drawing-chain. M—Bag. N—Hanging cage. O—Man who directs the machine. P, Q—Men emptying bags.

Fig. 5.3 Reversible Mining Hoist (Mid-16th c.)
Georgius Agricola (1556) p. 158

edge over competitors. But more generally it may have been due to his awareness of what the targeted readers and potential customers were interested in and could understand. Many of the incomplete depictions may nevertheless have conveyed all the information needed by carpenters or locksmiths who were experienced in the fabrication of such devices. As a rule, then, early modern images of machines could afford to be incomplete because they could presuppose knowledge of the real makers of machines; and they did so not always tacitly but sometimes invoke these practical experts' insight explicitly in the accompanying texts – from Vigevano in the fourteenth century, as we have seen, to Giuseppe Ceredi (ca. 1520 – ca. 1570) in the sixteenth century.[9]

This does not mean that besides potential customers the theaters of machines addressed the practitioners who fabricated machines as well. The price of these books alone would have deterred practitioners from purchasing them. Less splendidly outfitted and thus more easily affordable theaters appeared in the course of the seventeenth century, e.g. *Theatrum Machinarum* (1607–1614) by Heinrich Zeising (fl. 1610), which appeared in six small volumes, three of them in octavo. It should also be noted that the texts of most of the theaters were not written solely in Latin (with the exception of Besson) but also in one or sometimes two or more vernaculars. Ramelli, who composed his texts in Italian and French, omitted a Latin version.[10]

Almost all of the texts that accompanied the images in the theaters of machines explain the function of the depicted device. They often point to the savings the proposed device achieved in labor power and costs. Some texts also explain the device's kinematic path. Mechanical explanation of the working of the device or its parts occurs extremely seldom and usually amounts to some brief hints – we will return to this below.

The dominance of military engineers among the authors of the theaters of machines does not mean that the devices presented were predominantly those particularly employed in warfare. The main types of machines of the age – (a) hoists, cranes, and hauling devices; (b) ships, wagons, and other vehicles; (c) various kinds of mills; (d) pumps and other water lifting devices – did not pertain to one field of application only but were used in almost all of them: in military engineering, construction, mining, and the various civilian trades. This is why the devices presented in the theaters of machines belong to all of these fields of employment, and furthermore, why special books on fortification like *Delle fortificationi* by Buonaiuto Lorini (1545–1611) also present, besides gun mounts or temporary bridges and so on, hoists, pumps and other water lifting devices, as well as several kinds of mills.[11] And this also explains why one encounters significant sixteenth-century literature on mechanical engineering not only in these theaters but also in books on mining and particularly on architecture.

[9] See the quotation in Popplow (1998) 71.

[10] For mechanical engineers' awareness that craftsmen only had a limited ability to adapt machine proposals, see ibid. 74f.

[11] Buonaiuto Lorini: *Delle fortificationi: libri cinque, ne' quali si mostra con le piu facili regole la scienza con la pratica....* Venetia: G.A. Rampazetto, 1597.

That the early modern literature on architecture dealt with machines and mechanic engineering from the start in the mid-fifteenth century (Alberti and Filarete) was particularly because the canonical ancient text on architecture, Vitruvius' *De architectura libri decem*, discussed machines employed on construction sites and in quarries as well as surveying and other instruments. Readers of Vitruvius' work were also introduced to some basic statics, particular the law of the lever. The sixteenth-century editions of this work thus contain interpretations in words and pictures of these machines and devices. These interpretations are important contributions to the sixteenth-century literature on mechanical engineering.[12] (Fig. 5.4)

Fig. 5.4 Vitruvius' Archimedean Screw (1521)
Reconstructed by Cesare Cesariano – Cesariano (1969) f. clxxi[v]

[12] For Alberti and Filarete, see the chapter on Architecture. For interpretations of book X of Vitruvius, see in particular Cesare Cesariano: *Di Lucio Vitruvio Pollione de architectura libri decem*. Como: Gotardus de Ponte, 1521 and Daniele Barbaro: *I dieci libri dell' architettura di M. Vitruvio*. Venice: Franceschi and Chrieger, 1556. For Vitruvius' discussion of basic statics, see book X chap. 3.

5.4 Reasoning on Mechanics

As already indicated, apart from a few exceptions one looks in vain for explanations using mechanical concepts in the famous series of theaters of machines.

Ramelli and Branca almost always gave a description of the kinematic chain of the device at hand but never discussed why the machine parts involved interacted as they did. Besson and Errard contented themselves with some occasional indications of mechanical principles or possible mathematical reasons without clarifying exactly what was meant. The same holds for their occasional invocation of Archimedes. In Veranzio's theater one finds some indications of an odd understanding of mechanics – e.g., flyweights "increase" the force of the motor (*Magnam autem vim adunt Pondera*) – that may, however, prove very interesting on closer inspection. Only Zonca consistently explained the performance of a device at hand by referring to the principles of simple machines.[13]

The remarkable theoretical abstinence to be seen in the theaters of machines is most likely due to the expected grasp and interests of their addressees rather than to a lack of knowledge or interest on the part of the authors. Familiarity with the kinematic chain of presented devices is already clearly evident in a considerable number of fifteenth-century illuminated manuscripts, as well as knowledge of the characteristics of the simple machines used in this chain.

This is particularly clear in texts that gave advice about the fabrication of a device. Taccola, for example, when presenting a draw well powered by a treadwheel and, again, presenting a waterwheel driving a flour mill, advised a potential manufacturer in a way that shows his familiarity with the properties of the lever as incorporated in and governing the wheel and axle. More explicit examples can be found in Francesco di Giorgio Martini's *Trattati*. He opened the discussion on his huge series of flour mills by first explaining the characteristics of the wheel and axle, and even provided a formula for calculating the capacity of this simple machine. On the margin of the sheet he added a diagram of the wheel and axle captioned with the words "Reason of the lever" (*Ragione dela lieva*).[14]

As is well known, the investigations of machines and their parts, as well as the attempts at conceptualizing these devices undertaken by the Italian artist-engineer Leonardo da Vinci (1452–1519) went far beyond anything done by other engineers of the fifteenth or sixteenth centuries. However, as is also well known, these endeavors remained largely unknown to his contemporaries since Leonardo planned to compose a systematic treatise on the subject but did not realize this project. Many of Leonardo's manuscripts dealing with machines

---

[13] The texts of the theaters discussed here can easily be found in DMD. As to Besson and Errard see, for example, DMD IDs be14, be 17 and er01, er02, er27. For Veranzio's flyweights see DMD ID ve22. – The term "kinematic chain" is not, of course, an actors' term but was developed by Franz Reuleaux in the last quarter of the nineteenth century. – Six different "simple machines" have been identified since Antiquity: the lever, the inclined plane, the wedge, the screw, the wheel and axle, and the pulley.

[14] For Taccola, see DMD IDs tit05 and tit19; for Francesco di Giorgio Martini, see DMD IDs gm62a and gm62b. For a discussion of Francesco's text, see Long (2004) 122f. (Interestingly, Errard de Bar-le-Duc opened his theater of machines with a picture of a wheel and axle together with a laconic indication of its principle; see DMD ID er01). For Taccola and Francesco di Giorgio Martini, see also Galluzzi (1991) and (1996).

were published only centuries later. They therefore cannot be included in the corpus of early modern technological literature on mechanical engineering discussed here.[15]

Thus, one can state that, as some of the illuminated manuscripts of the fifteenth and sixteenth centuries unmistakably testify, the engineers of that time were familiar with basic mechanical conceptions of machines – their various kinematic chains and some principles of the simple machines. To all appearances, these engineers acquired this knowledge not by books but by oral instructions, cooperation, and travelling – and probably occasionally also from illuminated manuscripts by fellow engineers. In other words, it seems unnecessary to look for learned manuscripts or books as sources for their understanding of mechanics. As Stillman Drake put it:

> *Unlike the other [scholarly] traditions [of mechanics] [...] the technological tradition is timeless; it is a living and growing body of knowledge existing in all periods and transmitted largely by oral instruction to engineers, military men, builders, and artisans. Much of the technological tradition has always been accessible without recourse to written sources.*[16]

Up to the mid-sixteenth century, the science of mechanics – statics (science of weights), kinematics, and related parts of the Aristotelian physics of motion – was taught (intermittently) in traditional ways at (some) universities, that is, generally without any reference to the *artes mechanicae*, which were seen as mere crafts.[17] In the second half of the sixteenth century new impulses for studying these themes came from outside the universities, partly from educated men like Federico Commandino (1506–1575) or private scholars like Guidobaldo del Monte (1545–1607) but partly also from practical mathematicians like Giovanni Battista Benedetti (1530–1590) and Niccolò Tartaglia (1499–1557). However, their fresh approaches to statics and kinematics cannot be regarded as being stimulated, let alone enabled, by challenging objects or new achievements in contemporary mechanical engineering or by the illuminated manuscripts and other technological literature of the age. Rather, this fresh approach was stimulated by the rediscovery and study of classical literature on mechanics – the works of

[15] Besides innumerable sheets with drawings and texts on mechanical devices scattered among Leonardo's notebooks and portfolios, there is one codex which can be taken as a collection of notes and rough drafts for the planned work – the *Codex Madrid* (Madrid, Biblioteca Nacional Sign. 3986 u. 3987), first discovered in 1965 – Leonardo (1974).

[16] Drake & Drabkin (1969) 7.

[17] See Popplow (1998) 14–19. The short-lived revival of original studies in statics and kinematics by university scholars in the High Middle Ages — Jordanus' science of weights in the thirteenth century, the Oxford Calculators at Merton College in the fourteenth century – may, as Stillman Drake suggested (Drake & Drabkin (1969) 7), be seen against the background of increased interest in precise weighing in the context of minting and coin assaying. For an overview of learned mechanics in the Middle Ages and in the 15th and 16th centuries, see Clagett (1979) and Drake & Drabkin (1969).

Archimedes, Hero of Alexandria, and Pappus as well as the Ps.-Aristotlian *Quaestiones mechanicae*.[18]

> This[19] is clear in the case of Commandino, whose main contribution consisted in his critical translations and commented editions of classic works on mathematics and mechanics – notably those of Archimedes, Hero of Alexandria and the *Collectiones* of Pappus. This is also the case as regards Guidobaldo's *Mechanicorum liber* (1577), certainly the most stringent treatment of the simple machines since Antiquity. This work provided a rigorously geometrical (Euclidean-style) analysis and conceptualization of these machine elements – elements that were neither new nor challenging objects like some aspects of gunnery, which Tartaglia was trying to tackle. Accordingly, Guidobaldo's *Liber* was of significance for scientists like Galileo rather than for contemporaneous mechanical engineers.[20] The theoretical efforts in mechanics by Benedetti, although he was practically engaged in several engineering projects, remained largely unconnected with the state of contemporary mechanical engineering. The chapters of his *Diversarum speculationum mathematicorum et physicarum liber* that treat statics and kinematics consist mainly of his critical analyses of much-discussed theoretical texts at that time rather than of machines.[21]

We can thus state that this start of a new phase of the science of mechanics was triggered by newly accessible (discovered, translated, edited) ancient and medieval literature on mechanics rather than by developments in mechanical engineering. That is to say, this revival of original studies in mechanics made no change initially to the situation in which scientific and technological literature on machines developed independently of each other.

---

[18] Just as the contemporary literature on mechanical engineering played no part in the revival of learned statics, so, conversely the rediscovered classical literature on mechanics was of little significance for the mechanical engineers of the time. The practical mechanics reflected in this classical literature was simply outdated compared to that of the sixteenth century. See, for instance, White (1978) 90f.

[19] For the following, see Renn and Damerow (2012).

[20] It should be added that the *Liber* was already translated into Italian in 1581 and into German in 1629. This is noteworthy since, before the second half of the seventeenth century, mechanical engineers did not usually resort to a seriously mathematical or geometrical treatment of problems they faced. See Guidobaldo del Monte: *Mechanicorum liber*. Pisauris [Pesaro]: Apud Hironimus Concordiam, 1577; [Guidobaldo del Monte]: *Le mechaniche ... tradotte in volgare dal Filippo Pigafetta*. Venice: Francesco di Franceschi Sanes, 1581; [Guidobaldo del Monte] *Mechanischer Kunst-Kammer Erster Theil: Von Waag, Hebel, Scheiben, Haspel, Keyl und Schrauffen. Begreiffend die wahre Fundamenta aller Machination / verfasst und mit aussführlichen KupfferFiguren erkläret durch Danielem Mögling* .... Frankfurt: M. Merian, 1629.

[21] Giovanni Battista Benedetti: *Diversarum speculationum mathematicorum et physicarum liber*, Turin: Apud Hæredem Nicolai Beuilaquæ, 1585. For Benedetti's criticism of Tartaglia's adaptations from Jordanus in his *Quesiti e inventioni*, see chaps. 7 and 8, and for his criticism of the Aristotelian *Mechanical Problems*, see chaps. 10–24. See also Renn and Damerow (2012) chap. 6.

5.5 Machine Science and Early Modern Statics: A Mismatch

This statement needs some qualifications, for seemingly contradictory reasons. On the one hand, mechanical engineers occasionally, and in some cases even consistently, resorted to the statics of simple machines, though in a rather pragmatic way. On the other hand, the improvement in the rigor of statics achieved by this revival of Archimedean-type statics deepened the gap between theoretical and practical mechanics and proved to be the point of departure of an independent and separate development of both sides.

As we have seen above, the law of the lever and its meaning for an understanding of simple machines such as the wheel and axle had occasionally been invoked in illuminated manuscripts of the fifteenth century, e.g. by Francesco de Giorgio Martini. Similar and more elaborated and consistent considerations on statics can be found in writings on machines in the period immediately before and after Guidobaldo's *Mechanicorum liber*. Two examples may suffice.

The[22] physician Guiseppe Ceredi (ca.1520–1570) is known, if at all, for his design of an Archimedean Screw (cochlea) for which he obtained patents in Parma and other north Italian city states. The Archimedean Screw, a widespread water lifting device in Antiquity, was known in the West only from literary sources (Vitruvius) and apparently not yet built and employed up to the time when Ceredi published his *Tre discorsi sopra il modo d'alzar acque da' luoghi bassi* in 1567. Ceredi, who had rich experience in the field of water lifting devices, dismissed the reconstructions of the ancient cochlea proposed so far – that is, implicit or explicit interpretations of Vitruvius' text like those by Cesare Cesariano (see Fig. 5.4) and Daniele Barbaro – and based his own approach not on this text but on considerations of the power (water, labor power, other factors) that was to drive the cochlea, the strength of the materials, the weight of the water to be lifted, and the angle of lift. In doing this, Ceredi resorted to scientific statics as a resource of mechanical engineering – he mentioned that "certain writings of Hero, Pappus, Dionisidorus" as well as those of Euclid and Archimedes had guided his investigations. He stressed, however, that these theories could not be put into effective practice without tests and experiments with models.

The same approach to solving design problems in mechanical engineering can be observed in Ceredi's critical discussion of an age-old and ubiquitously used mechanical tool, namely the crank or, more precisely, its shape, which was usually curved at that time.[23]

---

[22] For the following, see Drake (1976).

[23] For Vitruvius' description of the cochlea, see his *De architectura* Book X chap. 4. For the Vitruvius interpretations of Cesariano and Barbaro, see, for instance, DMD IDs vice171v and viba463d. For cochlea proposals by Ceredi, see also DMD IDs cer69, cer70, cer76 and cer78; for his discussion of cranks, which were curved in the hope they would function like flywheels, see also DMD IDs cer54, cer57, cer58, cer59 and cer67. A translation of relevant passages of Ceredi's text into English is provided in Drake (1976). For Ceredi and ancient texts on statics, see Drake (1976) 58 and Traetta (2019); for Ceredi's experience of contemporary water lifting devices, see Zanetti (2017) chap. 8.

Vittorio Zonca's *Novo Teatro* (1607), our second example, differs from other such theaters not only in that it does not primarily present ideas or proposals for possible (and impossible) machines but also machinery employed in and around Zonca's native city Padua in that period. In our context, the *Novo Teatro* deserves attention because its descriptions and explanations use systematically static conceptions of the presented machines.

To give an example: In presenting a flour mill powered by a horse whim, Ramelli and Zonca both describe the kinematic chain through which the motion of the whim is transmitted to the millstone. In contrast to Ramelli, however, Zonca argues also for certain arrangements and measures explicitly derived from an understanding of the machine parts involved as simple machines.[24] But it is not at all clear whether these static considerations were borrowed from the statics of simple machines as rediscovered and elaborated at the end of the sixteenth century. Consider Zonca's description of another flour mill powered by a horse whim:

It seems to the practitioners (prattici) that the movement (movimento) of this machine must be very easy if the bar to which the horse walking around is tethered is elongated, because they say that this bar is like an arm of a balance, the center of which is the beam placed in the vertical that makes the wheel turn. And as Arist. in the Mech. states that this is the case, that the parts of the balance most distant from the center are faster (più veloce), as is evident to the senses (al senso), and thus more easily moved […].[25]

It is noteworthy that Zonca, when invoking the law of the lever, refers first to the knowledge of practitioners and to learned authorities – in this case not Archimedes but Aristotle – only for confirmation of the practitioners' view. This shows not only that early modern practitioners – certainly not ordinary carpenters or metalworkers but experienced machine builders – had no need of learned assistance as regards the basics of simple machines. It also reminds us that such basic knowledge of statics was a rather trivial part of these practitioners' experience and insight into highly intricate mechanisms like gear systems assembled from several simple machines. Such intricate mechanisms had been developed in Antiquity and the Arabic Middle Ages and also in the West, in the High Middle Ages and the early modern period. Examples are the many different gearings invented for mills (and depicted in the illuminated manuscripts of the fifteenth century, particularly in Francesco di Giorgio Martini's *Trattati*, or the intricate gearing systems developed by Filippo Brunelleschi for his hoists (see the chapter in the present volume on the science of architecture). The most impressive gearing mechanisms can be found in mechanical clocks and particularly astronomical clocks.

Mechanical clocks driven by weights seem to have been developed in the tenth or eleventh century; they appeared in the fourteenth century as public clocks installed on church towers.[26] Astronomical clocks seem to have been invented in Antiquity, as the famous

---

[24] See DMD IDs ra120 and zo028. For Zonca's *Teatro Novo*, see Poni (1985) and Keller (1988).

[25] Vittorio Zonca *Novo theatro* (above note 5) 34. This rough translation can be found in DMD ID zo033.

[26] The gearing of water clocks like that of Ctesibius (third century BCE), which Vitruvius described, was also very intricate. The most important difference between the gearing systems of water clocks

mechanism of Antikythera suggests. One of the eldest known astronomical clocks in the West was the *Astarium* built by the physician Giovanni Dondi (ca.1300–1388). An elaborate early modern discussion of the gearing of astronomical clocks can be found in *De rerum varietate libri xii* (1557) by Geronimo Cardano (1501–1576).[27]

At the beginning of this section, we stated that the improved rigor of statics achieved by the sixteenth-century revival of statics in the Archimedean way widened the gap between theoretical and practical mechanics and proved to be the point of departure of an independent and separate development of both sides. This needs some further explanations.

This revival consisted in a critical study of known and rediscovered texts on mechanics and particularly attempts at refining the proofs of laws of statics they contained. In other words, this revival was an internal event in theoretical mechanics: Some of the transmitted theories on statics – particularly those of Jordanus (*De ponderibus*) and Aristotle (*Quaestiones mechanicae*) – were criticized, rejected, or dismissed. Others – including that of Hero/Pappus on simple machines – were subjected to a strictly Archimedean-type treatment by referring to radically idealized "machines," namely, geometric representations of simple machines – "machines" without weight, without limited stability, without friction and so on. This opened a deep gap between theoretical mechanics and the mechanics of machines employed at the time and represented and described in the theaters of machines and other writings by mechanical engineers.

The size of this gap can be seen in the statics of Guidobaldo del Monte, who stressed that statics is a theory of equilibrium of weights, that is, about the conditions under which weights remain mutually at rest, and not a theory of their motion. In contrast, practical mechanics uses levers and other simple machines exactly as tools for transmitting motion.[28]

---

and mechanical clocks was that the latter contained a necessary escapement to ensure a uniform pace of the clock. It should be mentioned that due to their uniform pace, mechanical clocks marked equinoctial hours independently of astronomical observations; see, for instance, Dohrn-van Rossum (1992). Probably the eldest extant written document that allows glimpses of the mechanism of fourteenth-century mechanical clocks is the operating instructions of a church clock in Lucerne (Switzerland) of 1385 (see https://staatsarchiv.lu.ch/schaufenster/geschichten/turmuhr_1385). Throughout the entire early modern period efforts were made to improve the drive of mechanical clocks (weights, spring) and particularly the escapement (balance beam, pendulum, balance wheel).

[27] For the mechanism of Antikythera, see de Solla Price (1975). Giovanni Dondi: *Tractatus Astarii.* ms. D. 39 Biblioteca Capitolare di Padova, 1389 (modern edition: Emmanuel Poulle (ed.): *Tractatus Astrarii: Giovanni Dondi dall'Orologio*. Geneva: Droz, 2003). A slightly older astronomical clock was designed by the abbot Richard of Wallingford (1292–1336) and is described in his *Tractatus horologii astronomici* of 1327; see North (2005). For Cardano's discussion in his *De rerum varietate*, see Geronimo Cardanus: *Opera omnia*. Lyon 1663 (Stuttgart-Bad Cannstadt: Frommann 1966), vol. iii, 186ff.; the text can also be found in DMD IDs carv186, carv188, and carv189a-c.

[28] This explicit demarcation between equilibrium and motion in del Monte's *Mechanicorum liber* was attached to Prop. V in chapter 3 on the pulley by Filippo Pigafetta, the translator of the *liber* into Italian (*Le mechaniche* … (above note 20). Pigafetta referred to an argument between Guidobaldo and a critic about the deviations of real pulleys from ideal ones; see Meli (2006) 32ff.

A few years after Guidelbaldo's *Mechanicorum liber*, the well-known Dutch practical mathematician and engineer Simon Stevin (ca.1548–1620) made a similar attempt at treating statics in a strictly Archimedean way. In 1586 he published *De Beghinselen der Weegconst* and *De Weeghdaet*. The *first* book contains his attempt to establish a rigorously geometric (Euclidean style) basis for statics. (His famous proof of the law of equilibrium on an inclined plane can be found here.) The *second* book discusses and proposes ideal designs of several machine parts on the basis of the principles developed in the first book.[29]

When put into practice, such proposals for machine parts derived from geometric theories of the simple machines manifested rather than bridged the gap between theoretical statics and practical mechanics of machine builders. To give a famous example: In the 1590s, when Galileo was a professor at the university of Padua, he was consulted about new oars for Venetian galleys. The proposal he made derived from sound static considerations but proved to be inapplicable for the galleys in use.[30] In this case, the proposal was useless because it neglected the concrete situation in which the oars were to be used. The general reason for the limited utility of proposals derived from geometric statics was this neglect in statics of the properties of the materials the proposed devices were to be made of. In other words, the difference between geometric and material constructs was not taken into account.

> A contemporary of del Monte and Stevin, the military engineer Buonaiuto Lorini made the point when discussing the difference between a speculative mathematician (*mathematico speculativo*) and a practical mechanic (*meccanico pratico*). He stated that the demonstrations and proportions established by mathematicians regarding geometric entities separated from the material world (*separati dalla materia*), do not hold in practice because the mathematicians' concepts (*concetti mentali*) fail to recognize the given existing impediments (*impedimenti coniunti*) in relation to matter (*materia*). In contrast to a mathematician, it is therefore precisely the task of a guiding mechanic to consider those impediments – the more so as there are no reliable rules (*regola sicura*). He must thus combine knowledge of statics (*cognitione delle Matematiche*) with a wide experience of machines.[31]

For an assessment of this situation in the history of early modern mechanical engineering one has to take into consideration that issues like the stability, rigidity, and strength of materials used in machines or the estimation of friction were questions of practical experience (without a *regola sicura*). They remained so up to the eighteenth century when initial attempts of explanations and calculations were

---

[29] *De Beghinselen der Weegconst* was published in 1586 in a single volume together with *De Weeghdaet* (The Act of Weighing), *De Beghinselen des Waterwichts* (Principles of the Weight of Water = Principles of Hydrostatics), and an appendix. Stevin's estate contains an unfinished treatise on cogs and staves in which he tried to find optimal designs for elementary gearing parts, again on the basis of his statics. The same holds for his treatise on mills (*Van de Molens*) (also in his estate). (See *The Principal Works of Simon Stevin*, vol. v: *cogs and staves*: chap. iii, *on mills*: chap. vii.) We return to this draft on mills below.

[30] For details of this episode, see Valleriani (2010), chap. 4, 150ff.

[31] Buonaiuto Lorini: *Delle fortificationi libri cinque*. Venezia: Rampazetto, 1596, book v, 196. Lorini also pointed out that tests of proposed devices performed with small models do not necessarily ensure that they will work in practice.

undertaken on the basis of the theoretical mechanics developed during the seventeenth century.[32]

However, the reason why mechanic engineering could not profit a great deal from theoretical mechanics as represented by del Monte or Stevin was not just the mechanics' abstraction from the materiality of machines. Rather, it was actually this limitation of mechanics to statics that, as we saw in the case of del Monte, could even exclude motion as a subject of scientific mechanics. This was an extreme but in no way "illogical" case, as Stillman Drake put it.[33] But even when considering simple machines as transmitting motion, this mechanics excluded dynamic issues like the power that drives machines. Issues of dynamics were either dismissed outright (for example, by del Monte and Stevin) or discussed in critical analyses of Aristotle's physics (for example, by Benedetti). These critical analyses, although they triggered the beginnings of early modern dynamics with Galileo, had no relevance for contemporary mechanical engineering. Around 1600 even basic mechanical concepts such as impact, momentum, force or mechanical work had still not been developed – that is, exactly the dynamical concepts mechanical engineering most urgently needed. For machine builders had long been concerned with the key issue of how to estimate or assess the power or force (of human labor, animals, water, springs, or wind) that was to drive a machine.[34]

5.6 Measurement of Driving Forces I – Men as Driving Force[35]

In early modern mechanical engineering up to the seventeenth century, the question of the measurement of driving forces was usually asked in the form of how much force can be saved by a machine. The economical context of this question was always explicit, as the saved force was measured in the number of laborers (and thus costs for wages) saved.

[32] "Until the late seventeenth century no attempts were made to determine the amount of work done and the amount lost through friction, and the early studies failed to produce satisfactory results. Signal progress was made during the eighteenth century, and towards its close important results were delivered by Smeaton, Watt, and Evans." Usher (1929) 334.

[33] Stillman Drake regarded del Monte's demarcation of equilibrium and motion as "the ground on which pre-Galilean writers considered it impossible, and indeed illogical, even to attempt a science of dynamics." See Drake & Drabkin (1969) 300 n.31. Galileo, indeed, already differed on this issue from his mentor Guidobaldo in his early writings on mechanics; see Drake (1987) 35.

[34] See Wolff (1978) 255ff.

[35] In order to avoid anachronistic associations or assumptions, we use the term *driving force* and not *momentum*, *force*, *work*, or *energy*, depending on what kind of motor is at stake. Until the first decades of the nineteenth century, the historical actors primarily used the term "force" where we would use "work" or "living force/energy." See, for instance, the entry "Kraft" in Johann Heinrich Poppe (1803–16) Teil 3 (1806).

To give some examples: In his *De re metallica*, Georgius Agricola described a certain ventilation device as very useful since it was not driven by "men who require pay." In presenting a sawmill, Jacques Besson stressed that, "with just the movement of two workmen, as much work is done as by eight in the ordinary way." Fausto Veranzio presented a press, the "stone" (pressing weight) of which could be operated by one man and, due to a certain pulley, achieved the same effects as with several men. A very subtle example can be adduced from Geronimo Cardano's *De subtilitate* where he presented a winnow that saved costs since it could be operated by a day laborer instead of a skilled (costlier) worker.[36]

Obviously, such remarks on the driving force saved by machines proved to be a central concern of mechanical engineers and their customers. True, some of these statements might have merely been advertising, but some are certainly estimations based on experience – albeit very rough estimations that lack guidance as to how to construct energy-saving devices. Such estimations must nevertheless be taken as a method of measuring driving forces when taking into account the fact that among the driving forces of early modern machines, men were still as important as animals, wind, and water.

As the theaters of machines and other sixteenth- and seventeenth-century writings on machines testify, men served as driving forces not only for work machines such as spinning mills, sawmills, lathes, and so on (though work machines were increasingly driven also by water). Rather, up to the eighteenth century, all kinds of heavy machinery – pumps and water lifting devices, hoists, cranes, hauling devices, and so on – were not exclusively driven by animals, wind or water but still by men, too, such as men driving thread wheels. (Fig. 5.5)

It is consequently not surprising that eminent eighteenth-century engineers like John Theophilus Desaguliers (1683–1744) and John Smeaton (1724–1792) took the average performance of men, measured by lifting definite weights in a certain span of time, as the starting point for attempts at establishing a unit of measurement of forces that drove machines.[37] It should be mentioned in passing that the *horsepower* as a unit of measurement

[36]Agricola (1556) 163f.: "[...] machinam fabricari utilissimum est, & quòd nullo egeat uectiario, cui merces danda sit [...]" – see DMD IDag206; Jacques Besson (1578) heading of plate 13: "[...] per quam tantum operis exhibetur agitatione duorum operariorum, quantum octo possint vulgari ratione [...]" – see DMD IDbe13, see also IDbe15, be16 and be 27; Fausto Veranzio *Machinae novae* (1615) heading of plate 25: "Lapis autem suo pondere id efficiet, quod plures homines, suis viribus minimè possint" – see DMD IDve25; Cardano *De Subtilitate* (1554) (1966) III, 395f.: "[...] sufficere potest vtilesque geruli habentur, qui longè minore pretio operariis conducuntur" – see DMD IDcars395.

[37]In Smeaton's famous paper of 1759 on the *Natural Powers of Water and Wind to turn Mills*, one reads: "Desaguliers makes the utmost power of a man, when working so as to be able to hold it for some hours, to be equal to that of raising a hogshead of water ten feet high in a minute. [hogshead = 63 ale gallons] [...]/[...] when working at a mean rate, the 30 feet sail will be equal to the power of 18.3 men, or of 3 2/5 horses; reckoning 5 men to a horse; whereas the effect of the common Dutch sails, of the same length [...] will scarce be equal to the power of 10 men, or two horses.

That these computations are not merely speculative, but will hold good when applied to works in large, I have had an opportunity of verifying; for in a mill with the enlarged sails of 30 feet, applied to the crushing of rape seed, by means of two runners upon the edge, for making oil; I observed that when the sails made 11 turns per minute, in which case the velocity of the wind was about 13 feet in a second, [...] the runners then made 7 turns in a minute; whereas 2 horses, applied

Fig. 5.5 Mill Driven by Men (1615)
Fausto Veranzio (1968) table 23

of power, still in use today, was also established at that time by James Watt (1736–1819) who referred to experiences with horses driving pumps in mines when choosing this unit for the measurement of his steam engine's performance.[38]

In the course of the sixteenth and seventeenth centuries, a change can be observed regarding the balance between the four driving forces of machinery, which shifted in favor of wind and water without displacing men and animals. This was generally due to the expansion of the application of machines, particularly of work machines, in various fields of production. But it was specifically due to the development of more powerful pumps and water lifting devices needed for the drainage of mines as well as of land. At this time, the driving force for these heavy water lifting devices, which sometimes took the shape of machine systems, could only be provided by watermills and windmills and these, in turn, needed to be further developed for this task.

As regards machines and machine systems for the drainage of mines, even wind and water soon proved to be insufficient driving forces. This led to the search for new forces of this

to the same two runners, scarcely worked at the rate of 3 1/2 turns in the same time." John Smeaton: "An Experimental Enquiry concerning the Natural Powers of Water and Wind to turn Mills, and other Machines, depending on a Circular Motion". In: *The miscellaneous papers of John Smeaton [...] Comprising His Communications to the Royal Society*. London: Longman et al., 1814, 27–79, 74f.

[38] See Maul (1980) 282.

kind and attempts that eventually led to the invention of steam engines (see section 3 of the chapter on mining science in the present volume). As regards machines for the drainage of land, the Dutch draining or marsh mills merit particular attention, that is *windmills* that drove scoop-wheels or Archimedean screws for draining marsh land or polders. Although used since the Middle Ages for this purpose, these windmills were significantly improved in the early modern period.[39] In the course of the seventeenth century *watermills*, too, were developed and employed for lifting water on a hitherto unknown scale, not only for draining mines but also for prestigious projects like ambitious water art displays in royal park areas. The most famous big machine of this type is the *Machine de Marly* built in the 1680s for the water spectacles in the park of Versailles, which was hugely admired as a technical marvel by contemporaries but eventually proved to be a failure.[40] (Fig. 5.6)

5.7 Measurement of Driving Forces II –The Beginnings of a Theory of Machines

All these developments and improvements that brought forth more effective machinery were accomplished by mechanical engineers on the basis of proven solutions combined with (not always successful) alterations such as amplifications, refinements, and so on. In other words, these developments were undertaken by practitioners with no mechanical theory of machines at their disposal, as the mechanical theory of dynamics established in the seventeenth century – the century of Galileo, Descartes, Huygens, and Newton – had yet to be further developed for engineering use.

This situation becomes apparent when one compares books on mills published in the first decades of the eighteenth century with the representation of mills in theaters of machines of the period around 1600 or with Johann Faulhaber's (1580–1635) booklet on mills from 1617. Leonhard Christoph Sturm's (1669–1719) *Vollständige Mühlen-Baukunst* (1718) or Jacob Leupold's (1674–1727) *Theatrum machinarum molarium* (posthumously published in 1735), for instance, provide much more instructive descriptions of mills, showing the constructive improvements made in the seventeenth century in great detail. They also show,

[39] In the sixteenth century, the Dutch tower mill, that is a mill with a rotatable cap or roof, replaced the post mill in the Netherlands and northern German countries and was further developed to become the most advanced mill type in Europe. It was used as a flour mill both in the Netherlands and generally in other European countries. See Stokhuyzen (1962).

[40] To give an idea of the scope of this project: The *Machine de Marly* was built to provide the water needed for the water arts at Versailles by pumping it out from the river Seine. The distance between the park of Versailles and Marly-on-Seine is around 5 miles; the altitude difference is around 525 feet. The water was lifted by several arrays of pumps arranged one above the other (*in toto* 250 pumps). This pump system was driven by 14 very broad undershot waterwheels with a diameter of around 40 feet. The *Machine* never managed to deliver the projected service and proved highly prone to faults, so that about 60 mechanics were permanently employed for its maintenance. Its construction cost the incredible sum of four million *livres*. The *Machine* was finally dismounted and replaced by a new one at the beginning of the nineteenth century. For a contemporary description, see Bernard Forest de Bélidor: *Architecture hydraulique*. Vol. 2, Paris: Jombert, 1739, 195–203. See also, for example, Ergang (1911) and Brandstetter (2005) and (2008).

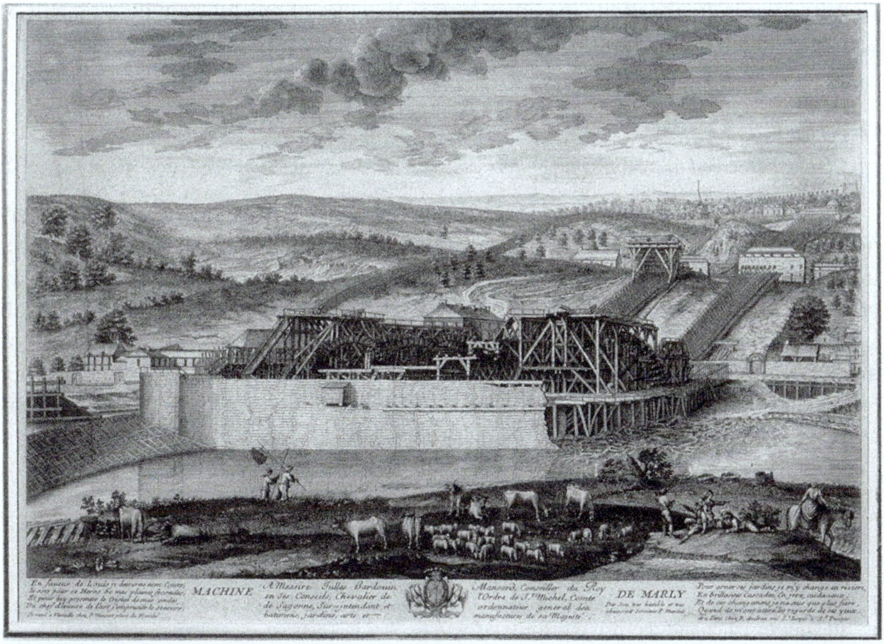

Fig. 5.6 Machine de Marly (1684)
Unknown seventeenth-century artist

however, that key performance parameters of mills were still not calculable.[41] The same holds for the famous Dutch books on mills published in the eighteenth century. To quote Robert James Forbes:

> *Strangely enough the standard works of the Dutch millwrights of that* [sc. 18th] *century contain nothing but structural details. Possibly corn millers were not so interested in the maximum output of their mills since they could organize their work, and the only early attempts at the calculation of windmill capacities concern drainage mills. We have plenty of eighteenth-century publications which claim to be "mathematical and mechanical calculations on drainage mills" but none were based on actual practical tests.*[42]

As the *Machine de Marly* dramatically proved, the discrepancy between effort and result in the field of advanced big machinery of the age called for new fundamentals of mechanical engineering. The artillerist (*Batterey-Meister*) Andreas Jungenickel (?-1654), author of a book titled *Clavis machinarum*, already saw

[41] Johann Faulhaber: *Ein Mathematische Newe Invention, Einer sehr nutzlichen und geschmeidigen Hauß‑ oder Handmühlin*, Augsburg 1616. Leonard Christoph Sturm: *Vollständige Mühlen-Baukunst,* Augsburg 1718; Jacob Leupold: *Theatrum machinarum molarium: Oder Schau-Platz der Mühlen-Bau-Kunst*. Ed. by Johann Matthias Beyer. Leipzig 1735.

[42] Forbes (1966b) 320. The standard works mentioned by Forbes are: Pieter Linperch, *Architectura mechanica of Moole-boek* (Amsterdam 1727); Johannes van Zijl, *Theatrum machinarum universal of Groot Algemeen Molen-boek* (Amsterdam 1734); Lendeert van Natrus, Jacob Polly, and Cornelis van Vuuren, *Groot volkomen moolenboek* (Amsterdam 1734/6).

alarming theoretical deficits in contemporary theaters of machines in the first half of the seventeenth century:

> *The majority of authors who wrote on mechanics, focused more on inventions and appliances (*wercke*) than on principles and fundamentals [...] if one would examine the fundamentals of [...] [those recent writings] one would find that the majority of machines or* wercke *described are wrong and cannot perform what they are supposed to perform [...]*[43]

Awareness of an urgent need for a new theoretical basis of mechanical engineering continued to grow among engineers around 1700. This can be seen, for instance, in Venterus Mandey's (1646–1702) and James Moxon's (1671–1708) *Mechanick-powers, Or, The Mistery of Nature and Art Unvail'd* (1696) and particularly in Jacob Leupold's *Kurtzer Entwurff* of 1725.[44] The crucial point for reaching a new level of mechanical engineering was the question of measuring and calculating the driving forces and the concomitants of this. Some brief remarks regarding the main problems facing such calculation may be in order here.

(1) Indirect measurement of driving forces.

One method of such calculations suggested itself, namely starting with calculating the required output of the machine. Perhaps the oldest paper that tried to calculate the performance of a complex machine dates back to the beginning of the seventeenth century – Simon Stevin's paper on marsh mills (*Van den Molen*). Instead of making the then futile attempt to calculate the driving force of winds, Stevin focused on calculating the mill's required performance, that is, the volume of water a scoop-wheel of given dimensions could lift with one revolution.[45]

Of course, the required output of a machine is rarely as easily calculable as in the case of lifting water. Moreover, the driving force of such a machine must not just surpass the

[43] "[...] daß der mehrentheils der Autores, so von der Mechanik geschrieben, mehr auff die Inventiones und Wercke gesehen, als auff die principia und fundamenta, daß kann ich also beweisen, daß, wann man das *Theatrum Machinarum* und die Schatzkammer Mechanischer Künste [...] nach den [...] fundamenten examiniren solte, würde man richtig befinden, daß der mehrentheil ihrer beschriebenen Maschinen, oder Wercke falsch sind, und nicht thun können, was sie thun sollen [...]."Andreas Jungenickel: *Clavis machinarum* (posthumously published in Nuremberg in 1661), 4. Jungenickel referred to Henricus Zeising's *Theatrum machinarum* (see note 5 above) and the German translation of Ramelli's *Le diverse et artificiose machine* (*Schatzkammer mechanischer Künste,* Leipzig 1620).

[44] Veterus Mandey and James Moxon: *Mechanick-powers, Or, The Mistery of Nature and Art Unvail'd*, London: Mandey, 1696; for Leupold's *Kurtzer Entwurff*, see section 3 of the chapter on mining science in the present volume.

[45] See Forbes (1966b) 311ff. Though not published at the time, Stevin's approach, and his calculation of the performance of scoop-wheels on the basis of his hydrostatics, was probably not unknown to contemporary millwrights (Stevin applied for several patents for improvements of mills) but found to all appearances few followers except for Jan Adriaanszoon Leeghwater's (1575–1650) *Haarlemmer-Meer-Boek* (1641). The draft was found in Stevin's estate, as mentioned above, and was revised by Jacobus Golius in 1634. This revised version was first published by J. Bierens de Haan in the nineteenth century: Simon Stevin: *"Van de Spiegeling der Singconst" et "Van de Molens". Deux traité inédits.* Ed. J. Bierens de Haan. Amsterdam, 1884.

weight of the load to be lifted by an increment; rather, the calculation must also take account of the power necessary to drive the moveable parts of the machine and, above all, to overcome friction.

(2) Friction and unequal rotational movement

The first theoretical treatment of friction in machines was published by Guillaume Amontons (1663–1705) in 1699. In the course of the eighteenth century the question of friction was further investigated by several mathematicians and natural philosophers, including Leonard Euler (1707–1783), Daniel Bernoulli (1700–1782), and Charles-Augustin Coulomb (1738–1806), who had command of Newtonian dynamics as well as the then new mathematical analysis.[46]

A machine's smooth transmission of motion depended on its gearing and particularly on cogs and the shape of their teeth. The efforts to find the optimal shape of these teeth undertaken in the seventeenth century by Simon Stevin and Ole Rømer (1644–1710) were resumed, for instance, by engineer-scientists such as John Smeaton, mathematicians like Philippe de la Hire (1640–1718), and natural philosophers like Euler in the eighteenth century.[47]

Other problems related to friction included vibration and unequal rotational movement. For equalizing the latter, *flywheels* and *flyweights* were in use in the West since the Middle Ages. The utility of these tools was well known; what remained unknown was their effect on the balance of forces before Newton established the concept of inertia movement at the end of the seventeenth century. On the basis of this new concept, in the eighteenth century, flywheels were investigated by mathematicians and natural philosophers, e.g. the Norwegian/Danish mathematician Jens Krafft (1720–1765).[48]

---

[46] Guillaume Amontons: *De la résistance* causée dans les machines, Mem. *de l'*Acad. Roy. Paris, 1699, 206–222. Leonhard Euler: *Sur le frottement des corps solides*. Mem. de l'Acad. Roy. Berlin, 4/1748, 122–132; and *Remarques sur l'effect du frottement dans l'équilibre*. Mem. de l'Acad. Roy. Berlin, 17/1762. Daniel Bernoulli: Commentatio de … *directione potentiarum frictionibus mechnicis adhibendarum* (1769) in Werke 3, 209–220. Charles-Augustin Coulomb: *Théorie des machines simples, en ayant égard au frottment de leurs parties et la roideur des cordages* (1781). Paris: De l'Imprimerie Royale, 1786.

[47] Simon Stevin's paper *On cogs and staves* was already mentioned above. In the 1670s, epicycloidal teeth of cogs were proposed by the Danish astronomer Ole Rømer; see, for instance, Forbes (1966a) 45. For Smeaton's cycloidal gears, see, for instance, Crosher (2014) 93. Philippe de la Hire: *Traité des Epicycloides et de leur Usage en les Machines* (1694), Mem. *de l'*Acad. Roy. Paris, 1730. Leonhard Euler: *De aptissima figura rotarum dentibus tribuenda*. Novi Comm. Acad. Scient. Petropol., 1760, 299–316.

[48] The effect of rotating wheels was known and used since time immemorial (e.g. the spinning wheel and potter's wheel). It was connected with the concept of impetus in antiquity and this connection continued in the West up to the seventeenth century; see Wolff (1978). In illuminated manuscripts on machines and in theaters of machines we frequently find the view that flywheels add to the driving force (from Francesco di Giorgio Martini up to Veranzio). As regards eighteenth-century texts on flywheels, some little-known German titles may be adduced: Jens Krafft: *Mechanica* (1773, German edition 1787) part II § 129; B.F. Mönnich: *Anleitung zur Anordnung und Berechnung der gebräuchlichsten Maschinen* (1779), part I/1, 68ff.; K. Chr. Langsdorf *Fortsetzung des Lehrbuchs zur Hydraulik welche eine Theorie der Schwungräder und ihre Anwendung bey Maschinen enthält* (1796); H.C. Brodreich: T*heorie des Schwungrads und der Kurbel* (1805). See Johann Heinrich Poppe (1999) 140f.

(3) Direct measurement of driving forces.

As regards waterwheels, the calculation of their driving force was in principle trivial in the case of *overshot waterwheels* (that is, waterwheels driven by the weight of water entering compartments at the top of the wheel) since weights as driving forces were a commonplace issue in statics. In contrast, such a calculation was a major challenge in the case of *undershot waterwheels* and *stream wheels* (that is, waterwheels driven by the impact or pressure of running water). Decisive steps towards a theory of fluid dynamics that allows such a calculation were made by Daniel Bernoulli and Giovanni Battista Venturi (1746–1822) in the eighteenth century. This was also decisive for evolving an aerodynamical understanding of the power that drives windmills.[49]

5.8 Conclusion

To sum up, we can state that in the field of machine science, scientific and technological literature developed largely independently of each other before the eighteenth century.

The medieval science of mechanics – if it existed at all apart from the short-lived heyday of Jordanus' *Scientia de ponderibus* and the kinematics of the Oxford calculators – largely ignored contemporary practical mechanics and the illuminated manuscripts on machines that emerged around 1400. And in turn, mediaeval practical mechanics neglected the Latin manuscripts on mechanical issues circulating in the distinguished sphere of the universities.

In the Renaissance, technological literature on mechanical engineering – illuminated manuscripts and, later, printed theaters of machines – found readers not only among potential customers such as princely military commanders but also among educated people like humanists. However, the revival of studies in classical theories of mechanics in the second half of the sixteenth century was not due to questions and problems posed by the contemporary technological literature on mechanical engineering but to the rediscovery of classical texts (Archimedes, Hero, and the Ps-Aristotelian *Quaestiones mechanicae*). The improved theories on statics that resulted from this revival found some resonance among engineers but proved of little help for several reasons, primarily because these theories excluded dynamic issues. We should add that this encounter of scientific and practical mechanics in the

[49] Daniel Bernoulli: *Hydrodynamica, sive de veribus et motibus fluidorum commentarii*. Strasbourg: Johannes Reinholdi Dulsecker, 1738; Giovanni Battista Venturi: *Recherches expérimentales sur le principe de la communication latérale du mouvement dans les fluides: appliqué à l'explication de différents phénomènes hydrauliques*. Paris: Houel et Ducros, 1797. The resistance of air had already become a challenging object for natural philosophers in the seventeenth century – e.g. Galileo, Edme Mariotte, Huygens, Newton; see section 6 of the chapter on gunnery in the present volume. John Smeaton's *Experimental Enquiry* of 1759 (see note 37 above) can be taken as the classic treatise on waterwheels as well as the windmill sails of his time.

years around 1600 entailed a remarkable ambivalence. On the one hand, it gave birth to the early modern figure of the engineer-scientist such as Stevin and Galileo, who scientifically investigated problems of mechanical engineering; and, on the other, it widened the gap between practical and theoretical mechanics even further, since the latter became a science of idealized (geometrical) subjects rather than material ones.

In the seventeenth century a certain stagnation can be observed as regards the technological literature on mechanical engineering. The descriptive and explanatory level attained by the theaters of machines created by Besson, Ramelli, and Zonca was rarely achieved by books on machines, let alone surpassed, before Leupold composed his series of such theaters at the beginning of the eighteenth century. This is all the more remarkable as the practical mechanics of the century – despite the havoc caused by the Thirty Years' War – was in no way stagnant, as proved by the development of drainage mills and other water lifting machines or even machine systems. However, these technical improvements were still achieved on the basis of rich practical experience and not through a better theoretical understanding of machines. The engineers' growing awareness of the limits of engineering on this basis was due particularly to the urgent need to calculate the performance of advanced and extended machinery.

The science of mechanics which developed basic concepts of modern (Newtonian) dynamics in the seventeenth century owed almost nothing to the contemporary technological literature on mechanical engineering. But this is not the full picture of the relationship between scientific and practical mechanics at the time. The latter constituted, in fact, an essential basis for the novel scientific method of studying mechanical questions by experimenting. To give a prominent example: the pendulum and inclined plane, both of which were known from and used in contemporary machines (Fig. 5.7), were the material means that allowed Galileo to perform his experiments in search of the law of falling bodies.[50]

Eventually, in the eighteenth century, the new science of mechanics, along with the new mathematical analysis, made it possible to tackle the urgent problems of mechanical engineering – measurement of the driving forces of machines, calculation of friction, finding optimal shapes of machine parts like cogs, and so on. This signaled the beginning of a new and close relationship between technological and scientific literature on mechanics which has continued to develop ever since.

[50] See Büttner (2008) and (2019); see also Lefèvre (2001).

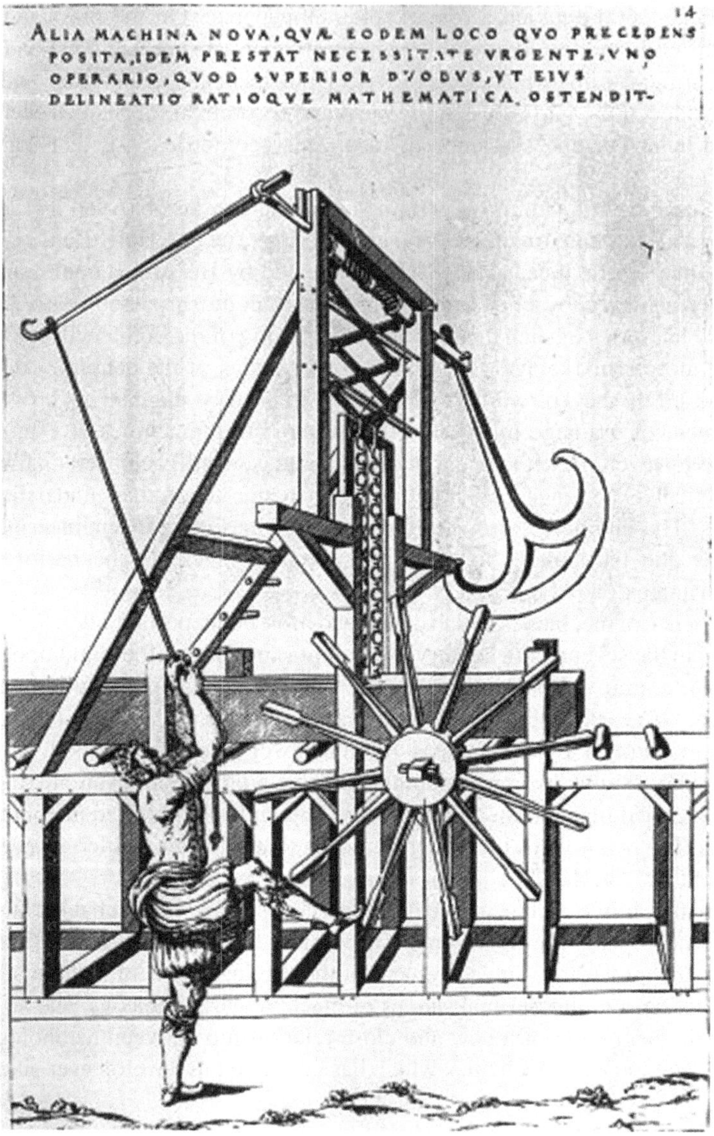

Fig. 5.7 Pendulum Flyweight (1578)
Jacobus Besson (2001) p. 14

References

Agricola, Georgius. 1556. *De re metallica*. Basel: Froben.

Besson, Jacobus. 2001. *Theatrum instrumentorum et machinarum (1578)*. Rome: Edizioni dell'Elefante.

Brandstetter, Thomas. 2005. "The Most Wonderful Piece of Machinery the World Can Boast of": The Water-Works at Marly, 1680–1830. *History and Technology* 21 (2): 205–220.

———. 2008. *Kräfte messen: Die Maschine von Marly und die Kultur der Technik*. Berlin: Kulturverlag Kadmos.

Büttner, Jochen. 2008. The Pendulum as a Challenging Object in Early-Modern Mechanics. In *Mechanics and Natural Philosophy before the Scientific Revolution*, ed. Walter Roy Laird and Sophie Roux, 223–237. Dordrecht: Springer.

———. 2019. *Swinging and Rolling: Unveiling Galileo's Unorthodox Path from a Challenging Problem to a New Science*. Dordrecht: Springer.

Cesariano, Cesare, ed. 1969. *Di Lucio Vitruvio Pollione de architectura libri dece (1521)*. Reprint München: Fink.

Clagett, Marshall. 1979. *The Science of Mechanics in the Middle Ages*. Madison: University of Wisconsin Press.

Crosher, William P. 2014. *A Gear Chronology: Significant Events and Dates Affecting Gear Development*. Bloomington: Xlibris.

de Solla Price, Derek J. 1975. *Gears from the Greeks: The Antikythera Mechanism – a Calendar Computer from ca. 80 B.C.* New York: Science History Publications.

DMD. 2006ff. *DMD – Database Machine Drawings – by Wolfgang Lefèvre and Marcus Popplow*. Berlin: Max Planck Institute for the History of Science. http://dmd.mpiwg-berlin.mpg.de/home.

Dohrn-van Rossum, Gerhard. 1992. *Die Geschichte der Stunde. Uhren und moderne Zeitordnung*. München: Hanser.

Drake, Stillman. 1976. An Agricultural Economist of the Late Renaissance. In *On Pre-Modern Technology and Science*, ed. Bert S. Hall and Delno C. West, 53–73. Malibu: Undena Publications.

———. 1987. *Galileo at Work: His Scientific Biography*. Chicago: University of Chicago Press.

Drake, Stillman, and I.E. Drabkin. 1969. *Mechanics in Sixteenth-Century Italy. Selections from Tartaglia, Benedetti, Guido Ubaldo & Galileo*. Madison: University of Wisconsin Press.

Ergang, Carl. 1911. Die Maschine von Marly. *Beiträge zur Geschichte der Technik und Industrie* 3: 131–146.

Forbes, Robert James. 1966a. Introduction [to chap. iii]. In *The principal Works of Simon Stevin*, ed. Robert James Forbes, 41–45. Amsterdam: Swets & Zeitlinger.

———. 1966b. Introduction [to chap. vii]. In *The principal Works of Simon Stevin*, ed. Robert James Forbes, 311–334. Amsterdam: Swets & Zeitlinger.

Galluzzi, Paolo., ed. 1991. *Prima di Leonardo. Cultura delle macchine a Siena nel Rinascimento*. Milan: Electa.

———., ed. 1996. *Renaissance Engineers, from Brunelleschi to Leonardo da Vinci*. Florence: Giunti.

Keller, Alexander G. 1988. [Review of] "Novo Teatro di Machine et Edificii 1607" by Vittorio Zonca, Carlo Poni. *Technology and Culture* 29 (2): 285–287.

Lefèvre, Wolfgang. 2001. Galileo Engineer – Art and Modern Science. In *Galileo in Context*, ed. Jürgen Renn, 11–27. Cambridge: Cambridge University Press.

———. 2002. Drawings in Ancient Treatises on Mechanics. In *Homo Faber: Studies on Nature, Technology, and Science at the Time of Pompeii*, ed. Jürgen Renn and Giuseppe Castagnetti, 109–120. Rome: Bretschneider.

———. 2003. The Limits of Pictures. Cognitive Functions of Images in Practical Mechanics – 1400–1600. In *The Power of Images in Early Modern Science*, ed. Wolfgang Lefèvre, Jürgen Renn, and Urs Schoepflin, 69–88. Basel: Birkhäuser.

———. 2010. *Picturing the World of Mining in the Renaissance – The Schwazer Bergbuch (1556)*. Berlin: MPI for the History of Science.

Leonardo da Vinci. 1974. *Codex Madrid I*. L. Reti et al. eds, Frankfurt am Main: Fischer.

Leng, Rainer. 2002. *Ars belli: Deutsche taktische und kriegstechnische Bilderhandschriften und Traktate im 15. und 16. Jahrhundert*. 2 vols. Wiesbaden: Reichert.

———. 2004. Social Character, Pictorial Style, and the Grammar of Technical Illustration in Craftsmen's Manuscripts in the Late Middle Ages. In *Picturing Machines: 1400–1700*, ed. Wolfgang Lefèvre, 85–111. Cambridge, MA: The MIT Press.

Long, Pamela O. 2004. Picturing the Machine: Francesco di Giorgio and Leonardo da Vinci in the 1490s. In *Picturing Machines – 1400–1700*, ed. Wolfgang Lefèvre, 117–141. Cambridge, MA: The MIT Press.

Maul, Kurt. 1980. Arbeit und Leistung. Ihre Bestimmung und Messung in der Technik seit dem 18. Jahrhundert. In *Technik-Geschichte: Historische Beiträge und neuere Ansätze*, ed. Ulrich Troitzsch and Gabriele Wohlauf, 269–301. Frankfurt am Main: Suhrkamp.

McGee, David. 2004. The Origins of Early Modern Machine Design. In *Picturing Machines: 1400–1700*, ed. Wolfgang Lefèvre, 53–84. Cambridge, MA: The MIT Press.

Meli, Domenico Bertoloni. 2006. *Thinking with Things: The Transformation of Mechanics in the Seventeenth Century*. Baltimore: The Johns Hopkins University Press.

Merrill, Elizabeth M. 2017. Pocket-Size Architectural Notebooks and the Codification of Practical Knowledge. In *The Structure of Practical Knowledge*, ed. Matteo Valleriani, 21–54. Cham: Springer.

———. forthcoming. The Spedale di Santa Maria della Scala and the Construction of Siena. In *Creating Place in Early Modern European Architecture*, ed. Elizabeth M. Merrill. Amsterdam: Amsterdam University Press.

North, John. 2005. *God's Clockmaker: Richard of Wallingford and the Invention of Time*. Oxford: Oxbow books.

Poni, Carlo. 1985. Scenari e fuori scena di un teatro di macchine. In *Vittorio Zonca: Novo teatro di machine et edificii: ix-liii*, ed. Carlo Poni. Milan: Edizione il Polifilo.

Poppe, Johann Heinrich. 1803–16. *Encyclopädie des gesammten Maschinenwesens, oder vollständiger Unterricht in der praktischen Mechanik und Maschinenlehre*. Leipzig: Voss.

Poppe, Johann Heinrich. 1999. *Geschichte der Technologie seit der Wiederherstellung der Wissenschaften bis ans Ende des achtzehnten Jahrhunderts (Göttingen1807)*. Vol. 1. Hildesheim: Olms.

Popplow, Marcus. 1996. Erfindungsschutz und Maschinenbücher: Etappen der Institutionalisierung technischen Wandels in der frühen Neuzeit. *Technikgeschichte* 63: 21–46.

———. 1998. *Neu, nützlich und erfindungsreich. Die Idealisierung von Technik in der frühen Neuzeit*. Münster/New York: Waxmann.

———. 2004. Why Draw Pictures of Machines? The Social Context of Early Modern Machine Drawings. In *Picturing Machines: 1400–1700*, ed. Wolfgang Lefèvre, 17–52. Cambridge, MA: The MIT Press.

Renn, Jürgen, and Peter Damerow. 2012. *The Equilibrium Controversy. Guidobaldo del Monte's Critical Notes on the Mechanics of Jordanus and Benedetti and their Historical and Conceptual Backgrounds*. Berlin: Edition Open Sources.

Stokhuyzen, Frederick. 1962. *The Dutch Windmill*. Bussum: C. A. J. van Dishoeck.

Traetta, Luigi. 2019. Giuseppe Ceredi. A Hydraulic Engineer in 16the-Century Italy. In *Explorations in the History and Heritage of Machines and Mechanisms*, ed. Baichun Zhang and Marco Ceccarelli, 17–27. Cham: Springer.

Usher, Abbot Payson. 1929. *A History of Mechanical Inventions*. New York: McGraw-Hill.

Valleriani, Matteo. 2010. *Galileo Engineer*. Dordrecht: Springer.

Veranzio, Fausto. 1968. *Machinae Novae (1615)*. Milano: Ferro Edizioni.

White, Lynn. 1978. *Medieval Religion and Technology – Collected Essays*. Berkeley/Los Angeles/London: University of California Press.

Wolff, Michael. 1978. *Geschichte der Impetustheorie: Untersuchungen zum Ursprung der klassischen Mechanik*. Frankfurt am Main: Suhrkamp.

Zanetti, Cristiano. 2017. *Janello Torriani and the Spanish Empire: A Vitruvian Artisan at the Dawn of the Scientific Revolution*. Leiden: Brill.

Chapter 6
Mining Science

6.1 Mining Science

Like other technological sciences, the science of mining comprises several kinds of knowledge – practical as well as theoretical – pertaining to the various components of this field of practice. These kinds of knowledge were not integrated by an overarching theory. Rather, they were united by the cooperation of several categories of actors: craftsmen, engineers, surveyors, and other experts. Extension and structure of the early modern science of mining was already outlined by the first comprehensive treatises on mining that appeared in the mid-sixteenth century – *Pirotechnia* by Vannoccio Biringuccio (1480–1538) and *De Re Metallica* by Georgius Agricola (1494–1555).[1]

> The maturity of both treatises is remarkable given the fact that their authors were unable to draw upon a tradition of literature on this subject. Works on mining of Antiquity had not been passed down to the West. And no literature had been produced by medieval miners, administrators of mines, physicians or scholars. It was not until around 1500 that a small booklet appeared on essential aspects of the mining industry. Written by the physician Rülein von Calw (1465–1523), it was plagiarized several times in the course of the sixteenth century. As a booklet, it was, however, too sparsely elaborated to be a model for Agricola's great treatise, though it might have been influential for the literary form of Agricola's first publication on mining, the *Bermannus* of 1530. For both publications present their content in the form of a dialogue – between a mining expert (*Bergverständiger*)

[1]Vannoccio Biringuccio: *De la Pirotechnia*. Venice: [Venturino Rufinelli], (posthumously) 1540; Georgius Agricola: *De Re Metallica Libri XII. Quibus Officia, Inſtrumenta, Machinæ, ac omnia denique ad Metallicam ſpectantia, non modo luculentiſſimè deſcribuntur, ſed & per effigies, ſuis locis inſertas, adiunctis Latinis, Germanicisque appelationibus ita ob oculos ponuntur, ut clarius tradi non poſſint*. Basel: Froben, 1556. Both treatises remained standard works well into the eighteenth century and were reissued several times and translated into various vernaculars. Biringuccio served the Republic of Siena as chief supervisor of the mines and also as the head of the mint. Georgius Agricola got acquainted with mining as physician of the mining district of Joachimsthal in Saxony (now Jáchymov, Czech Republic).

W. Lefèvre, *Minerva meets Vulcan: Scientific and Technological Literature – 1450-1750*, Archimedes 60, https://doi.org/10.1007/978-3-030-73085-7_6

and an apprentice (*Bergjunge*) in the case of Rülein and between two humanistic educated physicians and a mining expert in the case of Agricola.

The famous *Schwazer Bergbuch* was composed in 1556, the same year in which Agricola's *De re metallica* was published by Froben in Basel. However, being a memoir for the Imperial state administration in Vienna, it was not published at the time. The lavishly illustrated manuscript focuses more on administrative than technical issues.[2]

Despite differences regarding the division of the science of mining as well as the relative importance of the topics, Biringuccio's *Pirotechnia* and Agricola's *Re metallica* delineate the same principal extension and structure of the fields of knowledge that constitute this science: (1) prospecting and assessment of mineralogical yield, (2) mine surveying (*Markscheiden*), (3) construction of mining structures (pits/shafts and galleries), (4) extraction and transport of minerals, (5) the tools, appliances and machines employed, (6) processing minerals (particularly separation and smelting of metals). From the outset, early modern mining science also comprised administrative and legal matters, and often included knowledge about several mining districts.

Biringuccio deals with topics (1) to (5) at the beginning of the first book of his *Pirotechnia* whereas topic (6), particularly metallurgical matters, is presented in books 3 through 5 and additionally in book 7. He also deals with the extraction and processing of minerals other than metals, such as salt, sulfuric ores, alum, and others (book two).[3] In Agricola's *De re metallica*, book 3 is dedicated to topic (1), book 5 to topic (2), (3) and (4), book six to topic (5), and books 7 through 11 to topic (6). Book 12 deals with salts and glass production. Professional, administrative and legal matters are dealt with in books 1 and 2.

This principal structure of mining science remained unchanged from the sixteenth to the eighteenth century when, as we will see below, this science became a formal academic and educational enterprise. However, it must be added that, in the second half of the sixteenth and in the seventeenth centuries, no comparably comprehensive treatise was composed and published.

---

[2] From Greek Antiquity only the titles of works relating to mining are known; and as regards Roman Antiquity, one has to be content with some descriptions by Pliny the Elder; see Koch (1963) 3ff. Ulrich Rülein von Calw: *Ein nutzlich bergbuchleyn*, Augsburg 1505, reissued 1518, 1527, 1534, and 1539, as well as paraphrased, for example in Hans Rudhart's *Antzeigung des Nauenn Breythberuffen Berckwergks Sanct Joachimsthal* (Leipzig 1523), or even plagiarized, for example in the *Bergwerck und Probirbüchlein* (Frankfurt 1533) or in Georg Caspar Kirchmaier's *Institutiones metallicae* as late as 1687. For Rülein, see Darmstädter (1926) 13ff., Pieper (1955), Wilsdorf (1956) 181, Koch (1963) 21ff. Georgius Agricola: *Bermannus, sive de re metallica*, Basel: Froben, 1530. Around 1500 two other small booklets appeared on the then flourishing mining industry, the *Hymelisch Funtgrub* by Johann von Paltz (or Valtz) (1495) and *Judicium Jovis* by Paulus Niavis (c. 1495), which cannot be taken as models for Rülein, let alone for Agricola's *De re metallica*; see Koch (1963) 16ff. Of *Perkwerch* (= *Schwazer Bergbuch*) eleven manuscripts are extant; the most important are: Innsbruck Codex Dip(auliana) 856 (Landesmuseum Tyrol); Vienna Codex vindob(onensis) 10.852 (Austrian National Library), and Bochum Codex Sig. 3313 (Library of the Deutsches Bergbau-Museum).

[3] That is why his work is not discussed in detail here but in the chapter on chemistry in the present volume. Its descriptions of metalworking are discussed in the chapter on gunnery.

> Lazarus Ercker's (1528–1594) *Aula subterranea* (1573), certainly the most important sixteenth-century treatise on metallurgy after Biringuccio, did not investigate the other topics of mining. Of Georg Engelhardt Löhneyss' (1552–1622) *Bericht vom Bergwerck* (1617), apart from the illustrations of machines, only the detailed accounts on economic and administrative matters are of value, whereas all other topics are plagiarized from Ercker and Agricola. Not until the turn of the seventeenth century do we meet again with a comprehensive treatise on mining: Balthasar Roessler's (1605–1673) *Speculum metallurgiae* (1700), which confirms the principal structure of early modern mining science established in the mid-sixteenth century.[4]

However, as we shall see, in the period before 1700 texts on special aspects of mining appeared – such as surveying, machines, smelting and assaying etc. And we also come across literature on or in connection with mining in the early modern period that addressed aspects of the life and deeds of miners without being an integral part of the literature pertaining to mining science. In the sixteenth century, for example, a few collections of sermons were printed that used several aspects of the life and procedures of miners as similes for Christian dogmas or virtues.[5] These texts will not be discussed in the following. We will also not consider books that belong to the science of mining – texts on mining laws, for example, or reviews of mining districts – but do not deal with the technical aspects of mining.

The different kinds of knowledge pertaining to the principal topics of mining science range from craftsmen's knowledge to expert knowledge including elements of scientific knowledge. Accordingly, a variety of relationships between practical and scientific knowledge can be expected in the literature on mining. Focusing on ore or metal mining,[6] we will first look at mining literature concerned with the construction of mining structures as well as with extraction and transport of minerals, then turn towards literature about or pertaining to the tools, appliances and machines employed in mining, followed by literature on mine surveying (*Markscheiden*) and finally, literature pertaining to prospecting and mineralogy/geology. Metallurgical literature will be discussed not here but in the chapter on chemistry in the present volume.

---

[4] Lazarus Ercker: *[Aula Subterranea: d.i.] Beschreibung allerfürnemisten mineralischen Ertz- und Bergkwercksarten*, Prag 1573. Martin Stürtz' manuscript *Speculum metallorum* (1575) draws on the Schwazer Bergbuch when dealing with mining. Georg Engelhardt Löhneyss: *Bericht vom Bergwerck*, Zellerfeld: published by the author, 1617. Balthasar Roessler: *Speculum metallurgiae politissimum. Oder: Hell-polierter Berg-Bau-Spiegel. Wie man Bergwerck suchen, ausschürffen, dabey alles Gestein und Ertze gewinnen, fördern, rösten, schmeltzen soll.* (Postumously published Dresden: Winckler, 1700.

[5] The most popular of these collections is Johannes Mathesius' *Sarepta oder Bergpostill* of 1578. For *Sarepta* and other collections of sermons of this kind, see Koch (1963) 43f.

[6] As is well known, coal mining only gained importance with the beginning of the industrial revolution in the eighteenth century. The mining and production of non-metallic minerals such as rock salt, alum, flint, etc. did not inspire any literature comparable to that on ore mining.

6.2 Shafts, Galleries, and the Extraction of Minerals

The practical knowledge of digging mines was principally that of craftsmen, which was codified as little as that of other crafts.

> *The medieval miner could not draw on treatises on techniques and procedures of mining. In that age, mining was a matter of experience transmitted from one generation to the next, supported only by some written records or rules. This situation first changed with the turn of the Middle Ages and the early modern period.* (Manfred Koch)[7]

It is, indeed, remarkable that early modern treatises on mining even addressed technical matters such as the construction of shafts and galleries, breaking up rocks, and mining and producing desired minerals. But for whom were these treatises written? Were they read by miners? In fact, it seems that, up to the eighteenth century, mining operations of a craft character remained a matter of experience and rules transmitted from one generation of miners to the next without being codified.[8]

Biringuccio gave instructions on what to do and what to avoid for somebody planning to open and run a mining site. But he was very brief as regards the concrete procedures for the necessary tasks. Agricola, on the other hand, gave very detailed descriptions of the concrete work steps carried out in mining. Yet he did so in a book written in Latin.[9] Interestingly, the German edition of *De re metallica* did not sell as well as the original Latin edition.[10] One may well wonder for whom these descriptions had been written and published – for sovereigns and state officials, mining inspectors (*Berghauptmänner*) or pit foremen (*Steiger*), for possible investors or for physicians like Agricola himself? Anyway, these instructions and descriptions are highly significant, not only for present historians but also in view of their possible value for contemporaries interested in certain aspects of mining for scientific reasons.

> Agricola begins his descriptions with questions of how to plan, start, and continue to expand the web of shafts and galleries in such a way that it is integrated into the net of ore veins given at the site as well as the practical demands regarding transport of materials, drainage,

---

[7] "… auf ausgesprochen bergtechnische Abhandlungen konnte [der Bergmann des Mittelalters] nicht zurückgreifen. Lediglich durch diese oder jene kurze handschriftliche Aufzeichnung oder Vorschrift unterstützt, war der Bergbau durch all die Jahrhunderte hindurch Sache der reinen Empirie, von den Vorfahren den Nachkommen gewiesen. Erst an der Wende vom Mittelalter zur Neuzeit trat […] hier eine Wandlung ein." Koch (1963) 17.

[8] In the eighteenth century, special manuals on such technical subjects appeared. To mention just some titles: William Hardy: *The Miners Guide: or, Compleat Miner*, Sheffield 1748; Karl Friedrich von Boehmer: *Über die Grubenförderung*, Freyberg and Annaberg 1791; F.W. Dingelstedt: *Versuch einer Anleitung zur Grubenzimmer und -Mauerung für angehende Bergleute,* Schneeberg 1793. As we will see below, by that time mining science had begun to be taught at special academies and other educational institutions in some countries.

[9] See Biringuccio *Pirotechnia*, introduction to book I, and Agricola *De re metallica*, book V. The instructions and descriptions given in Balthasar Roessler's *Speculum metallurgiae* (1700) amount to a veritable manual of mining technology.

[10] Georg Agricola: *Vom Bergkwerck XII Bücher*. Basel: Froben, 1557; see Bartels (1992) 297f.

and ventilation. He even goes into details of the carpentry work needed to secure and fasten galleries and shafts. He also describes in detail the methods of beating off stones and ores, and reports which tools were used by miners for rocks and stones of different hardness. Additionally, he discusses advantages and risks of using fire for wearing down and breaking rocks. Blasting, that is, the use of explosives for breaking up a rock formation, was not yet employed in Agricola's time.[11]

The boom of ore mining, in particular silver mining, in the late fifteenth and the first two thirds of the sixteenth centuries in central Europe implied a technical "revolution" as a consequence of the new dimensions of deep-level mining.[12] Mining in such depths faced a multitude of new challenges, particularly as regards the drainage and ventilation of pits, pumps and other water lifting devices as well as ventilation systems. Such challenges could only be met by developing the machinery employed in mining. This leads us to our next topic: Machinery.

6.3 Mining Machinery

Our knowledge about the tools and machines employed in sixteenth-century mining is based as much on pictorial testimonies as on texts.[13] Most important in this respect are the detailed descriptions of these appliances that Agricola gives in book VI of his *De re metallica*, descriptions supported by an abundance of equally detailed woodcuts. A valuable supplement to this information is the illustrations in the *Schwazer Bergbuch* as well as the 16 plates with engravings attached to Löhneyss' *Bericht vom Bergwerck*.

Agricola's portrait of the various miners' implements (apart from equipment for separating, smelting and assaying of metals) ranges from simple tools (hammer, mallet, chisel, crowbar, etc.) and simple means of transport (baskets, vats, wheelbarrows, mining cars (*Hundte*), etc.) to various lifting devices – winch or windlass (*Haspel*), hoists driven by men, horse capstans (*Göpel*), and hoists driven by waterwheels – various kinds of pumps – bucket chains (*Becherwerke*), suction pumps, and Paternoster chain pumps (*Heinzel Künste*) – and ventilation devices. Of the pumps, only one was a real innovation of the time, namely a vertical arrangement and combination of a number of suction pumps by which water could be lifted from very deep levels as was required by underground mining in the mid-sixteenth century.[14] Several ventilation devices constitute the other important innovation spurred by the deep underground mining of the time. It goes without saying that, besides such genuine

[11] The blasting method came into use in the first half of the seventeenth century – in 1627 in the mines of Schemnitz (today Banská Štiavnica in central Slovakia) and in 1643 in the Freiberg mines in Saxony; see Baumgärtel (1963) 51. For blasting techniques of the eighteenth century, see Franz Xaver Baader: *Versuch einer Theorie der Sprengarbeit*, Freiberg 1792.

[12] In the mid-sixteenth century the mining district of Tyrol had the deepest mines ever heard of in Europe – see von Wolfstrigl-Wolfskron (1903) 195.

[13] See Lefèvre (2010).

[14] DMD ID ag185. Agricola states (Agricola (1912) 184) that this *Ehrenfriedersdorfer Radpumpe*, as he calls this pump, was invented only ten years previously, that is, around 1540. Tin mines existed at that time in Ehrenfriedersdorf (Saxony).

novelties, the mining machinery also comprised many improvements and adaptations of traditional machines. One such example is the evolution of a waterwheel into the reversible waterwheel employed in Joachimsthal as well as in Schwaz.[15] A flatrod system (*Stangenkunst*) which became an important element in machine systems of the seventeenth and eighteenth centuries seems to have been unknown to Agricola but appears in plate 10 of Löhneyss' treatise.

Agricola's descriptions address not only all the details and the function of a tool or machine at hand – sometimes even giving measurements of its parts – but also explain elaborately how it is handled and operated by the miners and, furthermore, its construction or implementation at the work site. But Agricola does not delve into assumptions or theories about the mechanical principles effective in a machine or deliberations about possible improvements derived from such principles. In this regard, his depiction of the mining machinery of his time conforms to the illuminated manuscripts or theaters of machines of the sixteenth century.[16] In the early modern period, machines became subjects of theoretical treatments only in the last third of the sixteenth century (see the chapter on mechanical engineering). However, as regards the machinery of mining in particular, such treatments seem to have begun significantly later, namely in the second half of the seventeenth century. Three famous or notorious attempts at improving mining machines merit brief description as indicative of such a scientific approach.

The well-known commitment of Gottfried Wilhelm Leibniz (1646–1716) in the mining district of the Harz region can be adduced first in this context. In the 1680s, he tried to deploy windmills as drive units for pumps – horizontal windmills as well as a vertical one he devised himself. He also planned to apply the pumps not only for draining the pits but also for reclaiming the used water for further use by lifting it back to ponds above the mines. Why his attempts failed is still debated – whether because of an inadequate conception or lack of practical experience on his part or obstructive behavior by the mining practitioners or for other reasons. However, his attempt is worth mentioning as an indication of the fact that mining machinery had become a scientific subject and a field of interactions between practical and theoretical mechanics in the second half of the seventeenth century.[17]

The Swedish scientist and engineer Christopher Polhem (born Polhammar, 1661–1751) epitomized this interaction some decades later. He is best known as the founder of the *Laborium mechanicum* in Stockholm and for his "mechanical alphabet" – a collection of wooden models of mechanisms for didactic purposes. In 1690 the Swedish king commissioned him to improve the mining operations of the

[15] DMD ID ag199 and sbb234.

[16] For the difference between Agricola's description and these theaters' presentation of machines, see the chapter on mechanical engineering in the present volume.

[17] An eighteenth-century discussion of Leibniz' attempt can be found in Henning Calvör: *Historisch-chronologische Nachricht und theoretische und practische Beschreibung des Maschinenwesens, und der Hülfsmittel bey dem Bergbau auf dem Oberharze*, Braunschweig 1763, 108f. For present-day assessments, see, for instance, Gottschalk (1982), Bartels (1992) 91f., Hirsch (2000) 130–186 passim, Fettweis (2004) 194–243, and Wellmer and Gottschalk (2010).

country and in this capacity, he faced the pressing problem of improving mine drainage by use of water lifting machines. He proposed and built the prototype of a new pump, called the "syphon machine," which tried to take advantage of the atmospheric pressure for driving its pistons. In our context, it is irrelevant whether or not this pump actually worked.[18] What is of interest here is the attempt at developing a technical innovation starting from a scientific concept – the concept of atmospheric pressure established in the first half of the seventeenth century by Evangelista Torricelli (1608–1647).

Another, and this time successful attempt at overcoming the insufficient exploitation of the accessible water by waterwheels as motors of pumps was undertaken by the artillerist Georg Winterschmidt (1722–1770) in the 1740s. Similar to an earlier invention of two French clerics, de la Deuille and Denisart, he developed a new drive unit for pumps, namely the water column machine.[19] This machine makes use of the hydrostatic principle known as Pascal's paradox.[20] The water column machine was also used for lifting devices (*Fahrkünste*) in mines and remained in use well into the nineteenth century in regions that lacked sufficient combustibles for the steam engine, which was then customarily employed in the mines of England and elsewhere.

The pressing problem of efficient pumps for mine drainage was also an important context for the early history of the steam engine. Proposals "to drive up water by fire" (Edward Somerset) can be traced back to the 1660s. Thomas Savery's well-known steam engine of 1699 was a pump, though it proved too weak for use in mines, and pumps were also among Denis Papin's inventions.[21]

Against this background it comes as no surprise that eighteenth-century publications on the machinery employed in mines reflect the achievements in scientific mechanics made since the seventeenth century. In 1725 the renowned engineer, instrument maker, and mining commissioner Jacob Leupold (1674–1727) published a short exposé on necessary measures to be taken for improving the machinery employed in mines. In this exposé Leupold emphasized the importance of assessing the existing machines on the basis of theoretical mechanics.

---

[18] An eighteenth-century discussion of Polhem's "syphon machine," can be found in Henning Calvör: *Historisch-chronologische Nachricht* (note 17 above) 136ff. See also Bartels (1992) 318ff. or Döring (2003) 26f.

[19] Winterschmidt wrote a treatise on his machine in 1761 which was published in Henning Calvör: *Historisch-chronologische Nachricht* (note 17 above) 158–190. For Winterschmidt and the water column machine, see Bartels (1992) 360ff. and Döring (2003) 26ff. For Deuille and Denisart, see Académie des Sciences (Paris) (ed.): *Machines et inventions approuvées par l'Académie Royale des Sciences depuis son établissement jusqu'à present*, 5(1735) 159ff.

[20] According to this "paradox," the vertical pressure of a liquid depends solely upon the density and height of the liquid and is independent of its volume and the shape of the container.

[21] Edward Somerset (1602–1667): *A century of … inventions*, London: Grismond, 1663, reissued several times in the eighteenth century. A similar proposal was published by Samuel Morland: *Elévation des eaux par tout sorte de machines*, Paris 1685 and 1700. See also Thomas Savery: *The miner's friend or an engine to raise water by fire*, London 1702; Denis Papin: *Nouvelle manière pour lever l'eau par la force du feu*, Kassel 1707.

> Leupold recommended in particular calculations on this theoretical basis – calculations of the power applied (here we should add a passing mention of his invention of a device for water measurements) as well as calculations of the actual performance of a machine compared to the performance that could be theoretically expected. This application of theoretical mechanics demands not only exact measurements but also accurate, that is, exactly scaled plans (*Risse*) of a machine and all its parts in order to compare the actual machine with one conforming to theoretical principles. It should be added that Leupold did not make use of the new mathematical analysis.[22]

Measurements and calculations of machines became keywords of eighteenth-century treatises on machines of the mining industry. In Henning Calvör's *Historisch-chronologische Nachricht*, for example, one finds tables with the results of experimental tests of pump performances. Nicolaus Poda (1723–1798) to give another example, discusses each machine according to a fixed scheme in his treatise on mining machines: a) relations of its main parts, b) its effect (performance), c) the power applied. He concludes with twenty-two tables for the calculation of machines.[23]

Given the emphasis Leupold placed on accurate orthogonal machine plans (*Risse*) it seems appropriate to take a quick glance at some drawings of mining machinery. To indicate the development that took place in the timespan between 1550 and 1800, it may suffice to contrast a machine drawing from the sixteenth century with an orthogonal plan of a machine from the eighteenth century. (Fig. 6.1).

6.4 Mine Surveying (*Markscheiden*)

For technical reasons as well as for economic or legal ones, mine surveying is an indispensable component of mining and, thus, a topic of mining science. The exact demarcation of the property rights of private investors in mining may even have been the main reason for mine surveying in the beginning.[24] In any case, with deep-level mining it became an essential technical prerequisite as well. The principal technical problem that demanded surveying arose out of the praxis of digging both pit/shaft and gallery independently before they met: How could the planners ensure that they would meet and how could they determine how much digging remained to be done in both digs on a given date? Agricola's *De re metallica* (second half of

---

[22] Jacob Leupold: *Kurtzer Entwurff, auf was Arth die Verbesserung des Machinen-Wesens bey den Bergwercken zu veranstalten*. Leipzig 1725. For Leupold's *Kurtzer Entwurff*, see Koch (1963) 95ff.

[23] Henning Calvör: *Historisch-chronologische* Nachricht (see note 17 above). Nicolaus Poda: *Kurzgefaßte Beschreibung der, bey dem Bergbau zu Schemnitz in Nieder-Hungarn, errichteten Maschinen, nebst xxii Tafeln zu derselben Berechnung; zum Gebrauch bey der Schemnitzer Bergschule*. Ignatz Born (ed.), Prag 1771. Poda, who taught mathematics and mechanics at the Mining Academy in Schemnitz, composed this book as a textbook for aspiring mining officials. For Calvör and Poda, see Koch (1963) 98f. and 103.

[24] See Bartels (1992) 122ff.

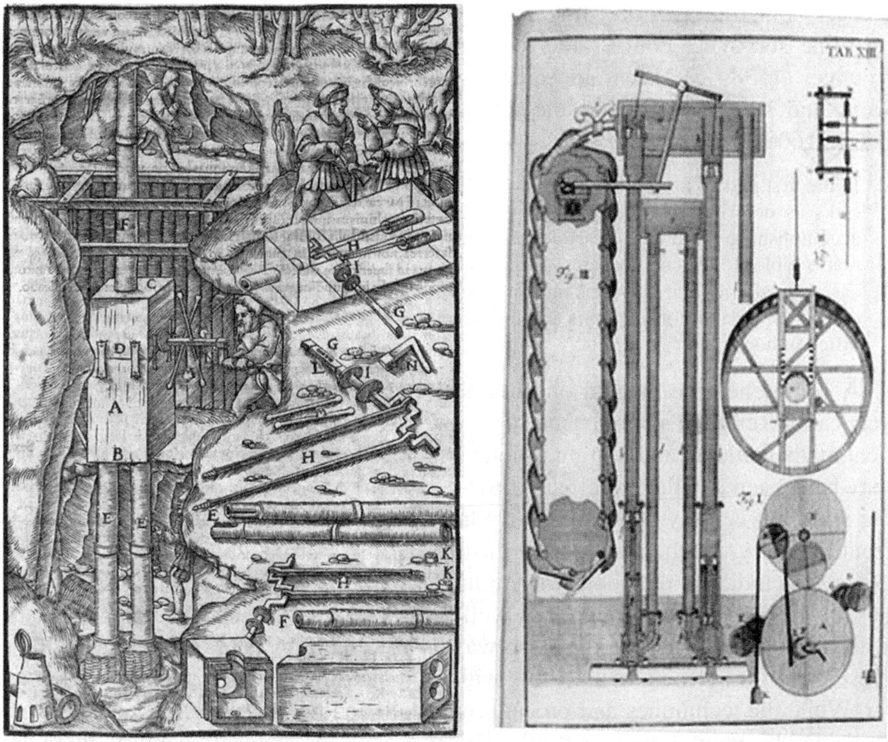

Fig. 6.1 Drawing Machines 1556 and 1763
Agricola (1556) VI, p. 139; Calvör (1763) table XIII

book V) provides detailed descriptions of techniques, procedures, and instruments applied and used by mine surveyors (*Markscheider*) in mid-sixteenth century mines.

> Agricola describes several surveying methods used by mine surveyors, all of which utilize the proportionality of the side lengths of similar triangles. That is, mine surveying was based on the same geometrical proposition as traditional surveying in land surveying, engineering projects, or in connection with siege and fortification. The particularity of mine surveying consisted in the methods of utilizing this geometrical rule under the circumstance of underground mining, that is, when specific places that could not be seen had to be sighted (indirect triangulation). A further challenge in mine surveying was the consideration of deviations of shafts and galleries from the straight line, which demanded angle measurement, leveling, and determining the linear extension of shaft or gallery from these measurements (by calculations or other means). Agricola also described the instruments of mine surveyors such as a compass, a quadrant (subdivided into 84 instead of 90 degrees), a levelling instrument, etc.

As with Agricola's other detailed descriptions, it is not clear whether those of surveying techniques were of use for contemporary practitioners or for the educated readers of his Latin work.[25] It is unlikely that they were used as a kind of textbook on mine surveying comparable to that by the physician Erasmus Reinhold the Younger (1538–1592) that appeared a quarter of a century after Agricola's major work and seems to be the sole textbook or manual on this subject published before 1600.[26]

> In the first part of his manual on mine surveying, Reinhold deals with the same surveying tasks as described by Agricola, providing concrete and very practical instructions for accomplishing these tasks. Besides explanations of surveying instruments, he offers the reader tables for calculations and diagrams. In the last part, he deals with different possible cases of galleries and ore veins meeting, an issue not discussed in Agricola. (Agricola discussed the position of ore veins and the structure of pits/shafts and galleries separately in different books.)

A comprehensive manual on mine surveying only appeared over one hundred years after Reinhold's book, namely the *Geometria subterranea* (1686) by the practical mathematician and mining official Nicolaus Voigtel (1658–1713).[27] This delay may have been a collateral effect of the decline of ore mining in the second half of the sixteenth century and particularly later, of the Thirty Years' War in the seventeenth century. As the two rapidly ensuing editions (1688 and 1693) of Voigtel's *Geometria* indicate, the book, written in German, was obviously a successful manual for aspiring mine surveyors as well as other mine officials. The manual deals with topics more or less in the same way as in Reinhold's book, although Voigtel's presentation is much more elaborate without impairing its principal didactic character. While the techniques and procedures taught and the instruments presented and explained by Voigtel are essentially the same as in the sixteenth century, there are notable developments in two regards.

First, the instruments – the compass and those for measuring horizontal and vertical angles – became more sophisticated and easier to handle than in Agricola's times. The suspended compass, for example, is presented by Voigtel but neither by

[25] It has even been suspected that Agricola mostly described methods that conformed to geometrical rules familiar to educated readers rather than those actually performed by mine surveyors of his time; see Morel (2020) 36f.

[26] *Vom Markscheiden kurtzer und gründlicher Unterricht* in: Erasmus Reinhold: *Gründlicher und warer Bericht vom Feldmessen [...] Desgleichen vom Markscheiden kurtzer und gründlicher Unterricht*, Erfurt 1574; it was reissued in 1597 and 1615. The book, an elaboration of an unfinished manuscript by his father, the renowned astronomer and mathematician Erasmus Reinhold the Elder (1511–1553), led to his appointment as a mining official (*Bergvogt*) of the mining district Saalfeld (Saxony). See Koch (1963) 40 and Bartels (1992) 122f.

[27] Nicolaus Voigtel: *Geometria Subterranea oder Marckscheide-Kunst: darinnen gelehret wird Wie auff Bergwercken alle Klüffte und Gänge in Grund und am Tag gebracht/ auch solche voneinander unterschieden werden sollen; so wohl Was bey Durchschlägen in Ersparung Kosten/ Bringung Wetters und Benehmung Wassers denen Zechen oder Gebäuden/ mit zu beobachten; Item/ Wie Streitigkeiten/ so sich unter miteinander schnürenden Gewercken offters zuereignen pflegen/ dem Maaße nach aus einander zusetzen; Sambt noch andern in nechstfolgendem Indice enthaltenen und zu dieser Kunst dienlichen Sachen; Allen Bergwercks-Liebenden zum Unterricht und versicherlichen Nutzen*. Eisleben: Dietzel, 1686.

Agricola or Reinhold.[28] Second, a new mathematical resource for surveyors is included in Voigtel's manual, namely sinus tables. This deserves attention in view of our investigation of the relations between technological and scientific literature in the early modern period. Voigtel did not confine himself only to explaining the technical terms (*sinus*, *cosinus*, etc.) and giving instructions on the use of trigonometrical functions for utilizing the geometrical measurements. He even explained how sinus tables are calculated, invoking Simon Stevin in passing – presumably referring to the German translation of Stevin's *Driehoukhandel*.[29] However, this does not mean that he explored the principles of trigonometry. Rather, he provided mine surveyors with a powerful means to make calculations without bothering them with theoretical issues.

Further improvements in the practical mathematics of mine surveying as well as its appliances were made in the eighteenth century.[30] But there was no principal turning point regarding the development of these fields. As to the instruments, their development potential on the basis of pre-industrial fabrication of mathematical instruments was exhausted in the eighteenth century. And as to mathematics, improvements were required not in theoretical mathematics but in practical mathematics, that is, mathematical procedures applicable by practitioners.[31] One development, however, could be regarded as a turning point: As we shall briefly address below, in the course of the eighteenth-century mining science became a science taught at educational institutions, and so also became the knowledge of mine surveying a teaching subject.

As in the preceding section on mine machinery, the present section on mine surveying should not conclude without taking a quick glance at graphical representations pertaining to this field, namely maps of mines constructed by mine surveyors.[32] For indicating the developments that took place in the timespan between 1600 and 1800, it may suffice to contrast a map/schema/section of a mine from the

[28] Ibid. plate 2, fig. 1. For compasses used by miners, see, for instance, Krause (1908), section II.

[29] Ibid. 34. Daniel Schwenter (ed.): *Simonis Stevini Kurtzer doch gründlicher Bericht von Calculation der Tabularum Sinuum, Tangentium und Seccantium*. Nürnberg: S. Halbmayer, 1628.

[30] The following eighteenth-century manuals on mine surveying are worth mentioning: August Beyer: *Gründlicher Unterricht vom Bergbau, nach Anleitung der Markscheider-Kunst*, Schneeberg 1749; Friedrich Wilhelm von Oppel: *Anleitung zur Markscheidekunst, nach ihren Anfangsgründen und Ausführungen kürztlich entworfen*, Dresden 1749; part II of Henning Calvör: *Historisch-chronologische Nachricht* (see note 17 above); Antoine François de Genssane: *La géométrie souterraine: ou traité de géométrie-pratique, appliqué a l'usage des travaux des mines*, Montpellier 1776; Johann Friedrich Lempe: *Gründliche Anleitung zur Markscheidekunst*, Leipzig 1782 and 1792.

[31] See, for instance, Johann Friedrich Lempe: *Bergmännisches Rechenbuch*, Freyberg 1787.

[32] To mention a prominent example of a cartographic work in connection with mine surveying: The mine surveyor Bernhard Ripkin (1682–1719) created the *Sylvae Hercinae Tabula* in 1715, that is, the first known map of an entire region with altitude readings, which served as the groundwork for the first printed map of the Harz mountains; see Bartels (1992) 318f. A short discussion and some recommendations concerning map constructing can be found in book 4, chap. 2 of Balthasar Roessler's *Speculam metallurgiae* (see note 4 above).

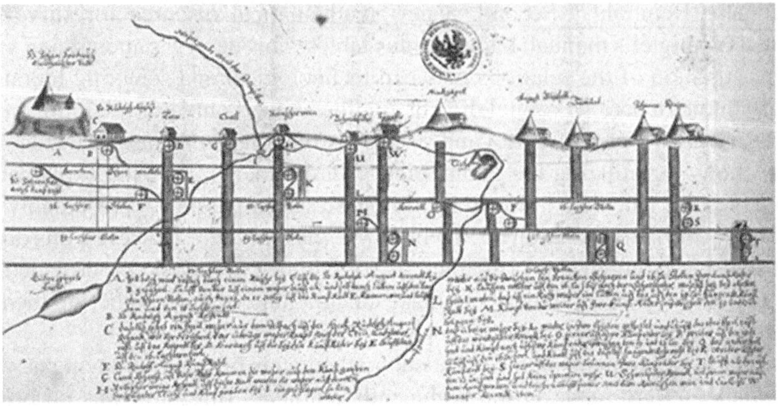

17th-c. schematic map of a mining district

Section through a mine (1785)

Fig. 6.2 Mining Maps – seventeenth and eighteenth-century
Unknown draftsman; Trebra (1785) – Lefèvre (2010) fig. 15 and 24

seventeenth century with a section through a mine from the eighteenth century. (Fig. 6.2).

6.5 Prospecting Minerals and Rocks

Prospection of metals or metal ores is carried out today with advanced techniques such as geophysical magnetic or gravity surveys. None of these techniques was known or available in the early modern period. In this period valuable minerals were found partly by chance if such minerals were exposed on the Earth's surface and partly by experienced people familiar with the signs that could indicate the presence of certain minerals in the underground. The knowledge of such signs was transmitted from one generation to the next in families – no apprenticeship, let alone schooling, existed for acquiring this knowledge.

However, some manuscripts on the topic circulated from at least the fourteenth century – they were called *Walenbücher* (Walloon books). These manuscripts, written by miners and prospectors mostly from the present-day Italian regions of the Alps, contained records of deposits and instructions for locating deposits of valuable minerals, particularly gold and precious stones, in certain regions – mostly in the Alps, but also in Saxony, the Harz region, Bohemia, and Poland. These manuscripts may have been valuable at the time as lists of deposits but can hardly be regarded as handbooks for prospecting.[33]

In the literature on mining from Agricola to Roessler, prospecting knowledge remained the knowledge of certain natural signs that might indicate the presence of ore deposits in the underground.[34] Moreover, despite the fact that roughly 150 years passed between the books by these two authors, the signs they listed are exactly the same: Foremost the occurrence of spring water (mountain water often runs off in veins of ores), furthermore the debris in the well, then exhalations detectable on

[33] See Koch (1963) 13. It is probably due to the promise prospectors saw in their (hardly reliable) listing of deposits that until well into the nineteenth century these manuscripts were copied again and again, and in the eighteenth century even in printed form. (The name *Walenbuch* is not derived from the name of the author of the eldest extant manuscript, an Antonius Wahle, but from a variant of the outdated German term *Welsch* for Mediterranean or Romanic.)

[34] As is well known, prospecting by dowsing rods (*Wünschelruten*) was in widespread use in the early modern period – up to the eighteenth century and even beyond. Accordingly, this method is discussed in the mining literature – very skeptically and ultimately dismissively by Biringuccio and Agricola, more ponderingly by Roessler. See Biringuccio (1959) 14; Agricola (1912) 38ff.; Roessler *Speculum metallurgiae,* book I chap. 31. Affirmative treatments of dowsing can be found, for example, in Nicolaus Solea: *Ein Büchlein vom Bergwerk, wie man dasselbige nach der Rutten und Witterung bawen soll*, n.p. 1600 or in Elias Montanus: *Bergwerckschatz, das ist ausführlicher und vollkommener Bericht von Bergwercken nach der Ruten und Witterung künstlich zu bawen*, Frankfurt 1618.; for these books, see von Klinckowstroem and Maltzahn (1931) 19, and Newman (2018) 75.

plants at a spot (for instance by the absence of white frost in autumn, by encrusted trunks or branches, etc.) and finally, certain striking patterns of vegetation.[35]

Besides these signs, experiences of adjacent concurrences of minerals played a role in prospecting. A place where ores of one metal had been found was seen as promising as regards ores of certain other metals; similarly, certain kinds of stone layers were known as potential receptacles of ore veins. The experiences and knowledge about the various ways in which veins of ores run and stretch within mountains was a further important resource not only for constructing shafts and galleries but also for prospecting. It comes as no surprise that Agricola regarded knowledge about kinds of soils, ordinary and precious stones, rocks, and metals, that is, knowledge of mineralogical and geological matters, as part and parcel of the professional expertise of an experienced miner.[36] In the context of our investigation it is therefore of interest to look at the relations that developed in the early modern period between mining literature and mineralogical and geological treatises.

As regards mineralogy, it is striking that Agricola, *the* outstanding mineralogist of the sixteenth century, did not deal with mineralogical issues in *De re metallica* but in some other treatises, particularly in his *De natura fossilium* of 1546.[37] Balthasar Roessler dedicated a few short chapters of his manual to mineralogical issues, distinguishing, for instance, between slate stones and sandy stones and exhibiting a basic classification of the former according to their colors.[38] One gets the impression that mineralogy, which owes a considerable part of its empirical knowledge to mining, played only a marginal role in the specific literature on mining before 1700.

Besides the classical literature on mineralogy before Agricola – Theophrastus, Avicenna, and Albertus Magnus – there was a long tradition of *Steinbücher* (lapidaries), that is, a tradition of manuscripts and booklets on stones, the focus of which was on precious stones as well as on magical properties ascribed to stones. They are valuable as catalogues of minerals known at a particular time and may have been of more significance for the mineralogical knowledge of miners than the classical works written in Latin. And this may also hold for the scientific literature on mineralogy of the sixteenth and seventeenth century, such as Agricola's book we have just mentioned, Konrad Gessner's treatise of 1565, or Bernardo Cesi's book of 1636.[39]

As regards geological issues, the courses of veins were an elaborately treated subject in all of the early modern manuals on mining – from Rülein von Calw's

---

[35] Agricola (1912) 37ff.; Biringuccio (1959) 15f.; Roessler *Speculum metallurgiae* (see note 4 above) book I chap. 27.

[36] *De re metallica*, book I, first paragraph.

[37] Agricola: *De natura fossilium libri X*, Basel 1546, English translation: Agricola (1955).

[38] *Speculum metallurgiae* (see note 4 above) chap I. 34ff.

[39] For *Steinbücher* see Koch (1963) 10ff; Konrad Gessner: *De omni rerum fossilium genere [...] libri*, Zurich 1565; Bernardo Cesi: *Mineralogia*, Lyon 1636. For early modern mineralogy, see Laudan (1987) chap. II; Oldroyd (1996) chap. I.

booklet to Agricola's *De re metallica* and Roessler's *Speculum metallurgiae*.[40] These elaborations presented a wealth of geological experience, often with a local character.

> To give a somewhat curious example: Agricola reported in *De re metallica* that miners preferred ore veins running in a west-east direction to those running in a north-south direction because they believed that the former yielded more than the latter. Agricola was critical not only of the miners' rationale for this difference (a presumed harmful influence of the sun on metals in rocks) but also regarding the generalization of this difference itself. Clearly, even if this difference held for the ore mines at a certain mining site it could not be generalized and extended to other mining districts without further information.[41]

This example shows the importance of gathering and accumulating knowledge about mining in several or, if possible, all known mining districts in order to establish general theories on ore veins in stone layers and mountains. It follows that surveys and reports on mining districts composed and published in the early modern period deserve attention in this context.[42] Together with manuals on mining such as Agricola's and Roessler's, they constituted an important basis for scientific geological works on ores in mountains that appeared in the eighteenth century. One example is Johann Gottlob Lehmann's (1719–1767) *Versuch einer Geschichte von Flötz-Gebürgen* (1756), which is regarded as one of the first scientific works on stratigraphy.

The early modern manuals on mining are not only of interest in geological terms because of the rich empirical knowledge on ores and rock formations they contain. Their discussions of this subject often assumed the character of true geological elaborations. However, as in the case of mineralogical matters, the most profound parts of Agricola's elaboration on geological subjects are not included in *De re metallica* but presented in the treatise *De ortu et causis subterraneorum*.[43]

> Presupposing, of course, God's primal creation, in this treatise Agricola discusses the natural origin of mountain waters, mountains and mountain regions, rocks, deposits, veins, earths, solid mixtures, stones, and finally ores, always examining the materials these items are made of as well as the natural agents and forces that produced them. In doing so he discusses critically the views held by the classical authorities (Aristotle, Theophrastus, Albertus Magnus) but also occasionally by contemporary (Paracelsian) chemists, before presenting his own views.

---

[40] Rülein von Calw: *Bergbüchlein* (see note 2 above) chap. 3 with 8 figures; Agricola: *De re metallica* book II with 21 figures; Roessler: *Speculum metallurgiae* (see note 4 above) book I chaps. 5–27 with 7 plates.

[41] Agricola adduced contradictory instances: see Agricola (1912) 73ff.

[42] To mention just some items of the early modern literature on mining districts in several countries: Peder Månsson's *Bergmanskonst*, a Swedish manuscript from the beginning of the sixteenth century; the accounts of mining locations in Sebastian Münster's *Cosmographia* (1544, reprinted many times); sixteenth-century *Bergchroniken* (mining chronicles) of several German mining towns and districts; Hans Bircherod's two books on gold mines in Europe, Asia, America and Africa (1689 and 1695): and as late as the eighteenth century, Franciscus Ernestus Brueckmann's *Unterirdische Schatzkammer aller Königreiche und Länder* (1727ff.) and Gabriel Jars' *Voyages Métallurgiques* (1774).

[43] *De ortu et causis subterraneorum libri V*, Basel: Froben 1546.

Agricola thus developed his own theory of mountains and minerals against the background of time-honored theories rather than against geological views of his time – if any such theories were developed in the mid-sixteenth century. In contrast, Roessler's few brief statements on geological issues written more than one hundred years after Agricola show that features of contemporary theories on geological issues or with geological implications left traces in manuals on mining.

In Roessler's *Speculum metallurgiae*, the chapter on *Würckung und Wachsthum der Ertz und Metallen* (effect and growth of ores and metals), for example, is clearly informed by Paracelsus' theory of matter (book I chap. 25). This may also apply to Roessler's contention that the vertical sequence of metals in a vein can be understood as a row of metal transformations (book I chap. 41).

Roessler's references to the Flood of the Old Testament as a cause for certain geological facts (e.g. book I chap. 28) deserve special attention. Invocations of the Flood as an explanation of such facts was far from unusual either in Roessler's time or in the preceding centuries. However, in the second half of the seventeenth century the Flood argument had become particularly important as a preferred answer to the challenges which new understandings of fossils presented to naturalists – preferred, for instance, by John Woodward (1665–1728) or Hans Jacob Scheuchzer (1672–1733).[44] Mining played a role in this development because it was due not least to findings of fossils made in the context of mining[45] that naturalists paid increasing attention to fossils of plants and animals in the seventeenth century – a focus that led to initial attempts at seeing the current state of the Earth as a result of radical changes in the past. The pathbreaking theory of fossilization developed by Nils Stensen (1638–1686) resulted not only in geological speculations like Woodward's but also in mechanistic hypotheses about the Earth's origin and formation.[46]

Independently of the question as to whether Roessler was actually aware of these developments in geology, the interplay between fossil records, geological theories and mining in his time seems to foreshadow such an interplay a century later that led to a breakthrough in stratigraphy. We are referring to the new, historical stratigraphy that took shape around 1800 and is worth a brief mention to conclude this section.

The English surveyor and engineer William Smith (1769–1839) was active both in mines and in channel digging projects that were undertaken at various sites in England, including mountainous ones, in the course of industrialization. This enabled him to observe, collect, record, and arrange a wealth of samples of strata as well as of fossils contained in them. This collection was certainly unique at the time as regards the territorial extension and

[44] John Woodward: *Essay toward a Natural History of the Earth*, London 1695; Johann Jacob Scheuchzer: *Herbarium Diluvianum*, Leiden 1723.

[45] For Agricola on fossils, see Agricola *De ortu et causis* (see note 43 above) III 39, *De natura fossilium* (see note 37 above) V 246f. and VII 327f. Leibniz' discussion of fossils in his *Protogea* (written 1700, published posthumously in 1768) related to specimens found in the mining district of Saalfeld (Saxony).

[46] See Laudan (1987) chap. II and Oldroyd (1996) chap. II. Niels Stensen: *De solido intra solidum naturaliter contento dissertationis prodomus*, Florence 1669.

depth of the strata the samples came from. Based on this record and arrangement, he established the "principle of faunal succession" that allows reconstruction of the original vertical order as well as the time sequence of strata. This was possible because, as he was the first to discover, each stratum contained a specific ensemble of fossil testimonies of living beings of the past. Beginning in 1799, he designed and published geological maps of various English districts and finally, in 1815, the first such map of England, Wales, and Scotland.[47]

6.6 Teaching in Mining Academies

As may have become clear by now, since Agricola's days mining science had become a complex science that comprised various fields of practical knowledge as well as fields of a scientific character or fields that presupposed proficiency in specific scientific domains such as mathematics, mechanics and so on, and thus demanded schooling. This is one reason why, in the eighteenth century, educational institutions for miners, particularly mining officials, were established in almost all continental European countries with silver mining districts. The cameralistic endeavors in several continental European countries should be seen as the general context of this development,[48] which began with *Bergschulen* (mining schools) at the beginning of the eighteenth century and culminated in the foundation of *Bergakademien* (mining academies) in the latter third of that century.[49]

The first mining school (*Bergschule*) was founded in 1716 in Joachimsthal (Saxony, today Czech Republic) and the first mining academies (*Bergakademien*) in Freiberg (Saxony) and Schemnitz (present-day Banská Štiavnica in central Slovakia) in 1765 and 1762/70

[47] Smith published his stratigraphic theory in a memoir together with this map (*A Memoir to the Map* […], London: John Cary, 1815) as well in two further books: *Strata identified by organized fossils*, London: W. Arding, 1816 and *Stratigraphical System of Organized Fossils*, London: E. Williams, 1817. For Smith see, for example, Eyles (1969) and Winchester (2001). Smith's achievement must be seen in the context of the high standard that scientific stratigraphy had reached in the eighteenth century, particularly in German states and Austria-Hungary, and furthermore, the stratigraphic explorations carried out by George Cuvier (1769–1832), who was also the leading expert in zoological morphology and thus for the assessment of fossils in this context.

[48] Emphasizing the role knowledge utilized in practice, the cameralistic doctrine saw the state authorities as the major promoter of useful knowledge and thus of the improvement of practice and the common good. Accordingly, the state authorities in several counties established "cameral science" as a university discipline for the academic education of state officials. This discipline comprised not only political and economic as well as fiscal and legal knowledge, rules of administration, etc. but also some mathematics and natural science.

[49] In contrast to early modern academies of science like the Royal Society in London or the Academie Royal des Sciences in Paris, these *Bergakademien* were not societies of scientists (learned societies, *Gelehrtengesellschaften*) but rather educational institutions dedicated to instructing mining experts and in particular aspiring mining officials. In a way they can be regarded as forerunners of the technical universities that emerged in the nineteenth century. See Klein (2016) and (2020).

respectively, followed by the mining academies in St. Petersburg (1773), Madrid (1777), and Paris (*École des mines* – 1783).[50]

In the context of our investigations two aspects of this development merit brief discussion at the end of this chapter: (1) In view of the question of the extension and structure of the mining science, the fields and topics taught at these educational institutions and (2) in view of the question of the interchange between technological and scientific fields of knowledge, the profile of the teachers as well as characteristics of the text books or manuals used or composed by them.

Questions relating to which topics and fields were to be taught and the teaching methods at these educational institutions were debated throughout the century. The debate gained momentum in the 1740s when the Saxon state official (*Kommissionsrat*) Carl Friedrich Zimmermann (1713–1747) argued for the establishment of an academy (learned society, *Gelehrtengesellschaft*) in which, in a first phase, savants from several scientific fields were to discuss which fields of knowledge constituted mining science before a second phase in which instruction of aspiring mining officials could commence.[51] In 1768 Johann Thaddäus Anton Peithner (1727–1792), professor for *Berg-Wercks-Wissenschaften* at the university of Prague, published a text, *Grundriß sammtlicher metallurgischen Wissenchaften* which can be taken as a representative result of the debates following Zimmermann's and subsequent proposals in around the mid-eighteenth century.

Peithner's lecture on mining science comprised, apart from knowledge concerning legal issues (part IV), practical knowledge of mining, mine surveying and assaying and smelting in the following order: Part I *Unterirdische Erdbeschreibung* (Geographia Subterranea), *Mineralienkunde* (Mineralogia), *Metallkunde* (Metallurgia); Part II *Unterirdische Meßkunst* (Geometria Subterranea), *Unterirdische Bewegungskunst* (Mechanica Subterranea), *Grubenbaukunst* (Architectura Subterranea); Part III *Probierkunst* (Ars Docimastica), *Schmelzhüttenkunst* (Ars Fusoria), *Berg-Fabriquen-Kunst* (Ars Fabricatoria Metallica).[52] The curricula actually taught at the mining schools and mining academies of the different mining districts were naturally not uniform. Whereas, for example, mineralogical and geological topics played a greater role at the mining academy in Freiberg than at that in Schemnitz, the latter, conversely, placed more emphasis than Freiberg on metallurgical topics, particularly in connection with the amalgamation process. In Freiberg, to mention just

[50] For the *Bergakademie* Freiberg, see Baumgärtel (1965) and Klein (2016) Part I, chap. 1, and for that in Schemnitz, see Konecny (2013) and Schleiff and Konecny (eds.) (2013). See also Weber (2015).

[51] Carl Friedrich Zimmermann: *Ober-Sächsische Berg-Academie, in welcher die Bergwercks-Wissenschaften nach ihren Grund-Wahrheiten untersucht, und nach ihrem Zusammenhange entworffen werden.* Dresden und Leipzig: Hekel, 1746, esp. pp. 26f. See Klein (2016) 17.

[52] Johann Thaddäus Anton Peithner: *Grundriß sammentlicher Metallurgischer Wissenschaften. In der Ordnung, nach welcher solche […] bey der Universität zu Prag von Thad. Peithner […] gelehret werden.* Prague: J.J. Clauser, 1768. Peithner's booklet *Erste Gründe der Bergwissenschaften aus denen physisch-metallurgischen Vorlesungen* (Prague: J.J. Clauser, 1770) contains only the paragraphs on the *Geographia subterranea*.

another interesting difference, special attention was paid to teaching drafting, that is the drawing of plans – ground plans, elevations, sections (*Grubenrisse*) – as well as maps.[53]

A characteristic feature of the training these mining schools and mining academies provided was the combination of theory and practice. The training comprised not only fields of knowledge with a scientific background but also instructions regarding the various practical tasks of mining.[54] Accordingly, we find even at the mining academies that the teachers included miners and mining officials with considerable practical experience as well as learned men proficient in sciences pertinent to fields of the mining science such as mathematics or mechanics, metallurgy and so on and, furthermore, hybrid experts who were both scientists and technologists.

To name just some personalities from the first decades of the *Bergakademien* in Freiberg and Schemnitz: The trained lawyers Friedrich Wilhelm von Oppel (1720–1769) and Christoph Traugott Delius (1728–1779), two leading figures among the teachers in Freiberg and Schemnitz respectively, acquired their expertise in mining matters as mining officials. Among the teachers for metallurgy and chemistry we find trained physicians: Nicolaus Joseph von Jacquin (1727–1817) and Giovanni Antonio Scopoli (1723–1788) in Schemnitz as well as the chemists Christlieb Ehregott Gellert (1713–1795) and Wilhelm August Lampadius (1772–1842) in Freiberg. Mathematics (including drafting) and mechanics was taught by practical mathematicians – Nicolaus Poda (1723–1798) and Johann Möhling (fl. 1800) in Schemnitz, and Johann Friedrich Wilhelm von Charpentier (1738–1805) and Johann Friedrich Lempe (1757–1801) in Freiberg. The first teacher of mineralogy in Freiberg, Christian Hieronymus Lommer (1741–1787), had been trained in assaying; his successor, the renowned mineralogist Abraham Gottlob Werner (1749–1817), had studied sciences at a university.

In accordance with the combination of theory and practice characteristic of the mining academies, manuals were composed not only for fields of knowledge with a scientific background but also for the instruction of knowledge pertaining to practical matters. The manuals and textbooks produced in connection with the instruction of the more theoretical disciplines comprised a broad spectrum ranging from short instructions in a how-to manner to true scientific books.

Delius and von Oppel published textbooks for the courses on *praktische Bergwissenschaft* (practical mining science). They focused on mine construction and other parts of mining technology, while metal smelting and processing as well as large parts of the natural sciences and mathematics were omitted.[55] Manuals on special fields of knowledge such as mine surveying or mechanics in connection with the machinery used in mining and smelting

---

[53] The mathematician Johann Friedrich von Charpentier (1738–1805), who taught mathematics and drawing at the *Bergakademie* Freiberg, published a *Petrographische Karte des Churfürstentums Sachsen* in 1778, that is, a geological map of the Electorate of Saxony.

[54] At the *Bergakademie* Freiberg, practical instructions, including exercises in pit and smelter, were given in the morning and the theoretical fields taught in the afternoon. In Schemnitz, the week was divided into days dedicated to practical instructions and others to teaching and learning theoretical topics. See Klein (2016) I chap. 1.

[55] Christoph Traugott Delius: *Anleitung zu der Bergbaukunst nach ihrer Theorie und Ausübung, nebst einer Abhandlung von den Grundsätzen der Berg = Kammeralwissenschaft für die Kaiserl. Königl. Schemnitzer Bergakademie entworfen*. Vienna: von Tratter, 1773. For Delius, see W. Arnold (1987). Friedrich Wilhelm von Oppel: *Bericht vom Bergbau*. Freiberg: Verlag der Bergakademie,

also had a predominantly practical character regardless of their use (and introduction) of some advanced mathematics.[56] These manuals do not differ essentially from those on similar topics composed in the decades before the rise of these educational institutions. The same cannot be said about the manuals on chemical, mineralogical and geological topics. Notwithstanding their didactic character, they do epitomize the type of technological literature that incorporated contemporaneous scientific knowledge on their subjects.[57] Finally, in the field of mineralogy/geology, we encounter books that resemble scientific treatises rather than textbooks.[58]

6.7 Conclusion

Mining science comprises, as we have seen, a number of fields of knowledge: construction, mechanical engineering, surveying, smelting, and prospecting as well as expert knowledge of minerals, ores, stones, and mountain formations; and we accordingly found various kinds of technological literature with particular relationships to sciences and scientific literature in each case.

It was certainly no surprise that the practical knowledge of digging, constructing, and fastening pits/shafts and galleries was as little codified as the knowledge of masons and carpenters in the case of architecture before the eighteenth century, and only entered texts then in connection with schooling. As regards mechanical engineering and surveying in the context of mining, we found in principle the same developments of the relationships of the respective technological literature to sciences (mechanics and mathematics) and scientific literature as we describe in the chapters on mechanical engineering and practical mathematics in the present volume. We should add, however, that the development of early modern machine

1769 and Leipzig: Crusius, 1772 (= a revision and extension of the manuscript *Ausführlicher und gründlicher Bericht vom Bergbau* (1740) by Johann Gottlieb Kern).

[56] Friedrich Wilhelm von Oppel: *Anleitung zur Markscheidekunst* (see note 30 above). Johann Friedrich Lempe: *Gründliche Anleitung zur Markscheidekunst* (note 30 above). Moehling: *Anleitung zur Markscheidekunst*. Vienna: "auf Kosten des Verfassers" 1793. Nicolaus Poda: *Kurzgefaßte Beschreibung der, bey dem Bergbau zu Schemnitz in Nieder-Hungarn, errichteten Maschinen* (note 23 above), Johann Friedrich Lempe: *Lehrbegriff der Maschinenlehre mit Rücksicht auf den Bergbau*, Leipzig 1795 and 1797.

[57] Christlieb Ehregott Gellert: *Anfangsgründe zur Metallurgischen Chimie: In einem theoretischen und practischen Theile nach einer in der Natur gegründeten Ordnung*, Leipzig 1750 and 1776. Wilhelm August Eberhard Lampadius: *Handbuch zur chemischen Analyse der Mineralkörper*. Freiberg: Craz 1801; ibid.: *Handbuch der allgemeinen Hüttenkunde, in theoretischer und praktischer Hinsicht.* 2 vols. in 5 parts, Göttingen: Dieterich, 1801–1810.

[58] Giovanni Antonio Scopoli's *Einleitung zur Kenntniß und Gebrauch der Foßilien. Füt die Studierenden* (Riga and Mietau: Johann Friedrich Hartknoch, 1769), for example, is a scientific classification of soils and minerals comparable to Abraham Gottlob Werner's *Kurze Klassifikation und Beschreibung der verschiedenen Gebirgsarten* (Dresden: Walther, 1787); Werner's most famous geological treatise (*Neue Theorie von der Entstehung der Gänge mit Anwendung auf den Bergbau*, Freiberg 1791) did not fail to emphasize topics of practical relevance for mining (see particularly chap. 9).

technology received special impulses from the need to drain and ventilate deep-level mines, and that some eighteenth-century proposals and inventions of more efficient pumps took scientific concepts as their point of departure, as did the proposals and inventions that finally led to the steam engine. And it should also be added that the literature on mine surveying generally equaled that of land surveying and remained largely unaffected by the developments in geodesy. (Literature on smelting is not discussed in the present chapter since metallurgy is a topic of the chapter on chemistry.)

The relationships between mining science and mineralogy, as well as the incipient geology of the early modern age, merit particular attention. The mineralogy inherited from Antiquity and the Middle Ages was the basis of early modern achievements in mineralogy from Agricola to Abraham Gottlob Werner – achievements, however, which owed almost everything to mining. And as regards geology, one can perhaps state that incipient early modern geology grew out of the stage of historical speculations triggered by a new understanding of fossilization, through empirically supported theories on the courses of ore veins, mountain formations, and particularly on strata of rock formations developed by mining experts like Johann Gottlieb Lehmann or William Smith. Here we come across a case of a scientific theory in which the ingredients developed for two hundred years in technological texts before they were systematized and represented in scientific treatises.

Finally, a brief remark seems appropriate regarding the textbooks written for and used in the mining schools and academies of the eighteenth century: It is important not to overlook that such educational material offered a suitable terrain for technological and scientific theories to come into contact and stimulate each other.

References

Agricola, Georgius. 1556. *De re metallica*. Basel: Froben.

———. 1912. *De re metallica: translated from the 1st Latin edition of 1556; with biographical introduction, annotations and appendices upon the development of mining methods, metallurgical processes, geology, mineralogy and mining law from the earliest times to the 16th century,* trans. and ed. Herbert Hoover and Lou Henry Hoover. London: The Mining Magazine.

———. 1955. *De natura fossilium*, trans. Mark Chance Bandy and Jean A. Bandy. New York: Geological Society of America.

Arnold, Werner. 1987. Christoph Traugott Delius (1728–1779) und seine Bedeutung für den europäischen Bergbau des 18. und 19. Jahrhunderts. In *Montanmedizin und Bergbauwissenschaften: 132–138*, ed. Wolfram Kaiser and Arina Völker. Halle (Saale): Martin-Luther-Universität Halle-Wittenberg.

Bartels, Christoph. 1992. *Vom frühneuzeitlichen Montangewerbe zur Bergbauindustrie: Erzbergbau im Oberharz 1635–1866*. Bochum: Dt. Bergbau-Museum.

Baumgärtel, Hans. 1963. *Bergbau und Absolutismus*. Leipzig: VEB Deutscher Verlag für Grundstoffindustrie.

———. 1965. *Vom Bergbüchlein zur Bergakademie. Zur Entstehung der Bergbauwissenschaften zwischen 1500 und 1750/70, Freiberger Forschungshefte D 50*. Leipzig: VEB Deutscher Verlag für Grundstoffindustrie.

Biringuccio, Vannoccio. 1959. *The Pirotechnia of Vannoccio Biringuccio*. New York: Basic Books.

Calvör, Henning. 1763. *Acta Historico-Chronologico-Mechanica*. Braunschweig: Waysenhaus-Buchhandlung.

Darmstädter, Ernst. 1926. *Georg Agricola*. München: Verlag der Münchner Drucke.
DMD. 2006ff. *DMD – Database Machine Drawings – by Wolfgang Lefèvre and Marcus Popplow*. Berlin: Max Planck Institute for the History of Science. http://dmd.mpiwg-berlin.mpg.de/home.
Döring, Mathias. 2003. Montane Energiegewinnung im Harz und Erzgebirge. In *Wasserhistorische Forschungen, Schwerpunkt Montanbereich*, ed. Christoph Ohlig, 21–46. Siegburg: Deutsche Wasserhistorische Gesellschaft.
Eyles, Joan. 1969. William Smith, Some Aspects of his Life and Work. In *Toward a History of Geology*, ed. C.J. Schneer, 142–158. Cambridge, MA: The MIT Press.
Fettweis, Günter B. 2004. *Zur Geschichte und Bedeutung von Bergbau und Bergbauwissenschaften*. Wien: Verl. der Österreichischen Akademie der Wissenschaften.
Gottschalk, Jürgen. 1982. Theorie und Praxis bei Leibniz im Bereich Technik, dargestellt am Beispiel der Wasserwirtschaft des Oberharzer Bergbaus. *Studia Leibnitiana, Supplementa* 22: 46–57.
Hirsch, Eike Christian. 2000. *Der berühmte Herr Leibniz*. München: C.H. Beck.
Klein, Ursula. 2016. *Nützliches Wissen. Die Erfindung der Technikwissenschaften*. Göttingen: Wallstein.
———. 2020. *Technoscience in History: Prussia 1750–1850*. Cambridge, MA: The MIT Press.
Koch, Manfred. 1963. *Geschichte und Entwicklung des bergmännischen Schrifttums*. Goslar: Hermann Hübener.
Konecny, Peter. 2013. Die montanistische Ausbildung in der Habsburgermonarchie, 1763–1848. In *Staat, Bergbau und Bergakademie – Montanexperten im 18. und frühen 19. Jahrhundert*, ed. Helmut Schleiff and Peter Konecny, 95–124. Stuttgart: Franz Steiner Verlag.
Krause, C. 1908. *Beiträge zur Geschichte der Instrumente in der Markscheidekunde*. Freiberg: by the author.
Laudan, Rachel. 1987. *From Mineralogy to Geology. The foundation of a Science, 1650–1830*. Chicago: The University of Chicago Press.
Lefèvre, Wolfgang. 2010. *Picturing the World of Mining in the Renaissance – The Schwazer Bergbuch (1556)*. Berlin: MPI for the History of Science.
Morel, Thomas. 2020. De Re Geometrica: Writing, Drawing, and Preaching Mathematics in Early Modern Mines. *Isis* 111 (1): 22–45.
Newman, William R. 2018. *Newton the Alchemist: Science, Enigma, and the Quest for Nature's "Secret Fire"*. Princeton University Press: Princeton.
Oldroyd, David. 1996. *Thinking about the Earth: A History of Ideas in Geology*. London: The Athlone Press.
Pieper, Wilhelm. 1955. *Ulrich Rülein von Calw und sein Bergbüchlein: mit Urtext-Faksimile und Übertragung des Bergbüchleins von etwa 1500*. Berlin: Akademie-Verlag.
Schleiff, Hartmut, and Peter Konecny, eds. 2013. *Staat, Bergbau und Bergakademie. Montanexperten im 18. und frühen 19. Jahrhundert*, Vierteljahrschrift für Sozial- und Wirtschaftsgeschichte Beiheft 223. Stuttgart: Steiner.
Trebra, Friedrich von. 1785. *Erfahrungen vom Inneren der Gebirge*. Dessau and Leipzig: Verlagskasse für Gelehrte u. Künstler.
von Klinckowstroem, Carl, and Rudolf von Maltzahn. 1931. *Handbuch der Wünschelrute. Geschichte, Wissenschaft, Anwendung*. München: Oldenbourg.
von Wolfstrigl-Wolfskron, Max Reichsritter. 1903. *Der Tiroler Erzbergbaue 1301–1665*. Innsbruck: Wagner.
Weber, Wolfhard. 2015. Erschließen, gewinnen, fördern: Bergbautechnik und Montanwissenschaften von den Anfängen bis zur Gründung Technischer Universitäten in Deutschland. In *Geschichte des deutschen Bergbaus*, ed. Wolfhard Weber, 217–408. Münster: Aschendorff.
Wellmer, Friedrich Wilhelm, and Jürgen Gottschalk. 2010. Leibniz' Scheitern im Oberharzer Silberbergbau. *Studia Leibnitiana* 42: 186–207.
Wilsdorf, Helmut. 1956. *Georg Agricola und seine Zeit*. Berlin: Dt. Verlag der Wissenschaften.
Winchester, Simon. 2001. *The Map That Changed the World: William Smith and the Birth of Modern Geology*. New York: Harper Collins.

Chapter 7
Practical Mathematics

7.1 Practical Mathematical Sciences

The literature pertaining to the various domains of early modern practical mathematics covers such a broad spectrum of subjects that it does not seem right to take it as the literature of one consistent and coherent technological field. That is why many historians prefer to speak of practical mathematical sciences only in the plural.[1] A quick glance at the class of early modern practical mathematicians already gives an idea of the broad spectrum of the fields of knowledge subsumed under the term practical mathematics.

This class comprised computing or reckoning masters (*Rechenmeister*), gaugers or calibrators (*Eichmeister*), and volume calculators (*Visiermeister*), for example of the volume of barrels, private tutors or school teachers of surveying and navigation, authors of calendars and ephemerides, and instrument makers (inventors as well as manufacturers). In addition, there were authors of textbooks on mathematics for practitioners (for merchants, gunners and fortification engineers, pilots, surveyors etc.) or of books on mathematical instruments (their conditions and use), and so on. Matthias Schemmel has given a rich and more precise description of this class:

> *The practical mathematicians attempted to solve practical problems in navigation, surveying, shipbuilding, fortification, gunnery, and similar fields of contemporary practical concerns. They handled engineering problems on commission, held lectures on practical issues, instructed seamen, and designed instruments. In view of this they may be considered practitioners. They were, however, distinct from the majority of practitioners by their reflection on the practical knowledge. They strived for mathematization of that knowledge, they edited ancient works on mathematics and translated them into vernacular to make them accessible to a broader readership […]. They were active largely outside the*

[1] See, for instance, Bennett (1987).

W. Lefèvre, *Minerva meets Vulcan: Scientific and Technological Literature – 1450-1750*, Archimedes 60, https://doi.org/10.1007/978-3-030-73085-7_7

universities and often relied on patronage to make their living, while institutional support grew only gradually.[2]

As regards the fields of practical mathematical knowledge, astronomy, surveying and navigation can be taken as the main domains in the early modern period. Each of these domains comprised many special fields of knowledge, and knowledge of specific instruments as well as particular segments of mathematics was associated with each of them. Furthermore, irrespective of their specific subjects, these main domains of practical mathematical knowledge were not isolated from each other. Rather, each of them drew upon the insights, results or procedures of the others. This is clear, for instance, in relation to the dependency of navigation and geodesy on each other and on astronomy, and conversely, the dependency of the latter on observational instruments in common use in all of the domains. It is therefore hardly surprising that we come across books which try to cover subjects from all of these domains, indicating that our classification of these various fields as practical mathematical ones is not at all anachronistic.[3]

This web of interrelations between the domains of practical mathematics allows us to discuss the various and seemingly heterogeneous literature on practical mathematical subjects by just following certain of these conjunctions between navigational, astronomical, and geodetic fields of knowledge. For pragmatic reasons we will concentrate largely on geodesy in this chapter, which is structured in terms of its manifold relations to the other domains. We will start with literature on practical geodesy, that is, surveying, with its instruments and mathematics, adding an excursus about literature on mathematics for practitioners. Focusing on literature on instruments and methods of angular measurement, we will touch on the domains of practical astronomy and cartography in connection with navigation. The chapter will conclude with a consideration of literature pertaining to the field of higher geodesy.

7.2 Surveying

Surveying is an age-old practice, older than classical Greek, Hellenistic or Roman Antiquity which continued and developed surveying procedures and rules employed in ancient Mesopotamia and Egypt. Like the practice of surveying, Mesopotamian and Egyptian literature on surveying led the way – particularly as regards area calculation. In the West, the Middle Ages and the early modern period could only build on a small fragment of this tradition, which was transmitted through texts by

---

[2] Schemmel (2008) 15.

[3] See, for instance, Daniel Santbech: *Problematum astronomicorum et geometricorum sectiones septem*. Basel: Petri & Pernam, 1561, or John Dee's famous preface to the first English edition of Euclid's *Elements*: *The elements of geometrie of the most auncient philosopher Euclide of Megara*. London: Daye, 1570. For the historical actors' view of the shared identity of the practical mathematical fields, see, for instance, Bennett (1986).

the Roman *Agrimensores*.[4] Early modern authors of books on surveying did, in fact, build on this tradition. The principal scope as well as the didactic style of these books show a remarkable resemblance to the inherited literary model.

> The core of these texts consisted of a few standard problems concerning area calculations of various geometric figures, dividing given areas into such figures, and indirect measuring – e.g. finding the distance between two inaccessible spots – as well as rules and strategies for how to solve them. The geometric laws that the solution strategies implicitly took advantage of were basically those of similar triangles, preferably right-angle triangles – not because of the Pythagorean theorem but rather because they facilitated area calculations. This core of surveying treatises can be traced back to Hero of Alexandria, whose *Dioptra* was formative for the Roman treatises. We find this core in the medieval descendants of the latter and also in early modern treatises on surveying, for instance in Jakob Köbel's *Vom Ursprung* (1522) or Martin Grosgebawer's *Vom Feldmessen* (1596).[5]

Many of the early modern treatises on surveying[6] also contain instructions in elementary geometric terms (definitions of a line, a triangle, etc.) and in basic calculating operations. With this, they shed a light on a fundamental social and cultural problem that beset early modern surveying up to the eighteenth century: the lack of practical surveyors with sufficient mathematical training. In the early modern period, beginning in the sixteenth century, there was a dramatic increase of surveying tasks – in connection with hydraulic construction projects, mine drainage, fortification/siege strategies, drastic redistributions of land, etc. – and, thus, a rising demand for surveyors which could not be met in societies that had not yet evolved a general educational system.[7] Complaints about incompetent surveyors (some of whom were actually charlatans) were a perennial topic of early modern treatises on surveying.

> in 1582, the English practical mathematician Edward Worsop published a treatise with the title *A discouerie of sundrie errours and faults daily committed by lande-meaters, ignorant of arithmetike and geometrie, to the damage, and preiudice of many her Maiesties subiects with manifest proofe that none ought to be admitted to that function, but the learned practisioners of those sciences*. Similar, though less drastic assessments of practical surveyors' mathematical abilities can be found as late as the end of the eighteenth century. J.G. Geissler, for instance, the editor and translator of the German edition of *Geometrical*

[4] For the Roman *Agrimensores* or *Gromatici*, see, for instance, Dilke (1971); for the writings of the Roman *Agrimensores*, see Campbell (2000).

[5] For the implicit mathematics in the texts of the *Agrimensores*, see, for instance, Folkerts (2014). For the formative role of Hero's *Dioptra* for the Roman texts on surveying, see, for instance, M. Cantor (1875). Jakob Köbel: *Von Ursprung der Teilug / Maß / vn Messung deß Ertrichs der Ecker*, Oppenheim, 1522 and republished as part 1 of his frequently republished *Geometrey* of 1535; Martin Grosgebawer: *Vom Feldmessen*, Schmalkalden: Schmück, 1596.

[6] For treatises on mine surveying, see Sect. 6.4 of the chapter on mining science in the present volume. Unlike treatises on common surveying, their focus lay on calculations of distances rather than of areas.

[7] As mentioned in the chapter on architecture, as early as the sixteenth century the shortage of competent surveyors in the dukedom of Milan led to the foundation of the first educational institution for architects in the West, the *Collegio degli Architetti, Ingegneri e Agrimensori* founded in 1563.

> *and Graphical Essays* by George Adams Jr., wrote: "[…] the ideas of the practical surveyor must be pressingly corrected because he is still struggling with wrong concepts which he, lacking the needed knowledge, is unable to amend."[8]

Besides practical instructions, the early modern literature on surveying written for practical surveyors provided rules (empirical formulas or recipes) for calculating distances and areas derived from learned geometry without going into detail or teaching this geometry. Even when introducing the nomenclature of learned geometry, it was teaching terms, not concepts or definitions. Learned geometry on the one hand and, on the other, the application of the rules derived from it, existed in different worlds. This literature for practitioners did not deal with the derivation of these rules; nor did learned geometry receive any stimulus from it. There was no exchange whatsoever between these worlds. But there were mediators between these worlds – the practical mathematicians.

Surveying was not the only practical field where mathematical techniques were needed. Beginning in the late Middle Ages, such techniques were needed in an increasing number of practical fields – commerce, warfare, architecture, mining, assaying, and so on. It was this situation that prompted the establishment of the first schools for elementary mathematics, most famously the *Abacus Schools* in Italy, and that brought forth the figure of the early modern practical mathematician and, furthermore, a special literary genre produced by this figure, namely books or booklets on mathematics for practitioners.

7.3 Excursus: Books on Mathematics for Practitioners[9]

The early modern books and booklets we will briefly consider here endeavored to provide practitioners with needed or desirable mathematical knowledge and were usually written in a country's vernacular. Interestingly, we also find books of a similar didactic character and with largely the same spectrum of contents written in Latin, that is, books that were certainly not aimed at normal practitioners.[10] The demand for such knowledge was obviously more general and the audience for this

---

[8] "… praktische[n] Vermesser, dessen Ideen doch vorzüglich berichtiget werden sollten, da dieser immer noch zu sehr mit fehlerhaften Begriffen in seiner Kunst zu kämpfen hat, die er aus Mangel näherer Kenntnisse nicht berichtigen kann." Adams (1795) XVII. Edward Worsop: *A discouerie of sundrie errours* […] London: Henrie Middleton 1582.

[9] An extensive list of early modern books on mathematics for practitioners can be found at https://de.wikisource.org/wiki/Rechenb%C3%BCcher

[10] To mention just a few of these Latin books on practical mathematics from the sixteenth century: Heinrich Stromer von Auerbach: *Algorithmus linealis* … (Leipzig 1504); Oronce Finé: *Protomathesis* (Paris 1532); Gemma Frisius: *Arithmeticae practicae methodus facilis* (Antwerp 1539); Sebastian Münster: *Rudimenta mathematica* (book 1, Basel 1551); Christopher Clavius: *Geometria practica* (Rome 1604).

literature wider than just the circle of certain practical professions such as surveyors or merchants.

The circle of the authors of this literature was as diverse as that of its audience. Not a few of them were scholars or men with a scholarly education: university teachers like Petrus Apianus (1495–1552) and Daniel Schwenter (1585–1636), physicians like Robert Recorde (1510–1558), theologians like Charles De Bouelles (1479–1567), and scientists like Christopher Clavius (1538–1612) and Johannes Kepler (1571–1630). The numerous authors without a scholarly education who composed such books included not only practical mathematicians in the narrow sense of this term, that is, computing or reckoning masters (*Rechenmeister*), gaugers or calibrators (*Eichmeister*), or volume calculators (*Visiermeister*), but also engravers like Edward Cocker (1631–1676), instrument makers like Leonard Digges (c. 1515 – c. 1559), cartographers like Johannes Werner (1468–1522), surveyors like Aaron Rathborne (1572–1618), fortification engineers like Benjamin Bramer (1588–1652), and so on. And among the authors of this category we even find translators and editors of classic works such as Euclid's *Elements* or some of the works of Archimedes and Hero of Alexandria: e.g. the practical mathematician and engineer Niccoló Tartaglia (1499–1557), who translated works of Euclid and Archimedes into Italian, or the architect Giovan Battista Aleotti (1545–1636) who translated Hero's *Pneumatica* into Italian.

The oldest Western books on *arithmetic* for practitioners must be seen against the background of the aforementioned *Abacus Schools* in Italy.[11] These schools, established and run by local authorities as well as private tutors since the thirteenth century, placed special emphasis on arithmetic in a commercial context. They tried to meet the needs of merchants related to bookkeeping and other commercial transactions. During the Middle Ages, copies of the Latin manuscript *Liber Abaci* (1202) by Leonardo Fibonacci (Leonardo da Pisa, ca. 1170 – ca. 1250) served as a kind of background textbook for instruction in *Abacus Schools*, as did the later *Arte dell'Abbaco* (also known as *Treviso Arithmetic*) anonymously authored in the Venetian vernacular and printed in 1478.

Fibonacci's *Liber* taught basic arithmetical operations including those with fractions, squares and square roots, dealt with the rule of the three, the conversion of coins (currencies) and various measures, and the solution of other problems pertaining to the daily business of commerce. Although the *Liber* also contained more advanced arithmetic and algebra, with regard to practical mathematics in the early modern period its most momentous innovation was probably the introduction of the Hindu-Arabic numeral system, that is, the place value system on which our basic arithmetical algorithms have rested since then. The *Treviso Arithmetic*, written in Italian, draws on Fibonacci's *Liber* and was significant for making this basic stock of practical arithmetic accessible to a broad audience.[12]

[11] For this background see, for instance, the "General Introduction" in Van Egmond (1981) and Goldthwaite (1972).

[12] Leonardo Fibonacci: *Il liber abbaci*, Baldassarre Boncompagni (ed.), Roma: Tipografia delle scienze matematiche e fisiche, 1857; L.E. Sigler (translator and ed.): *Fibonacci's Liber abaci: A translation into modern English* ..., New York, Berlin, Heidelberg: Springer, 2003. Digitization of the *Treviso Arithmetic*: https://web.archive.org/web/20120206013105/http://www.republicaveneta.com/doc/abaco.pdf. A translation of the text is published in Swetz (1989). For Fibonacci, see, for instance, Vogel (1971) and for Fibonacci's significance for fifteenth-century practical arithmetic, see Høyrup (2019), chaps. 13–16, 28, and 29.

Mathematics for commercial transactions became a standard feature of early modern books on practical mathematics. The Franciscan friar Luca Pacioli (1447–1517) dedicated one part of his seminal textbook *Summa de arithmetica* (1594) to the description of the commercial bookkeeping method, known as the double-entry accounting system. Other authors of books on practical mathematics followed this example, e.g. Heinrich Schreiber (Grammateus) (1495–1525) in his *Ayn new kunstlich buech* (1518), while Petrus Apianus published an entire book on commercial mathematics in 1527.[13]

As regards mathematics for surveyors, we have already mentioned that almost all treatises on surveying and mine surveying provided short instructions in basic mathematics. The arithmetic taught in these treatises was – apart from special commercial issues – essentially the same as that for merchants. However, these instructions often failed to cover the whole spectrum of arithmetic procedures for practitioners described in computational books of the time.

Popular sixteenth-century computing books (*Rechenbücher*) like that of Adam Ries (1492–1559) or Robert Recorde taught computing with the abacus or counting boards as well as with Hindu-Arabic numbers, integrals and fractions, sometimes also with squares and square roots, and also dealt with elementary algebra. Some of the sixteenth-century computing books provided more advanced arithmetic, going beyond what practitioners such as surveyors would or could acquire, e.g. *Die Cos* by Christoph Rudolff (ca. 1500 – ca. 1543) or the Latin *Arithmetica integra* by Michael Stiffel (1487–1567). These books already reflect the beginnings of a Western algebra we shall not discuss in detail here. Rather, we will confine ourselves to remarking that the practical mathematician Niccolò Tartaglia was among the famous pioneers of Western algebra.[14]

The instructions in basic mathematics provided by treatises on surveying and mine surveying naturally concerned, above all, *geometrical* issues. As already mentioned, they taught rules (empirical formulas or recipes) for dividing and calculating areas without entering into the geometrical theory these rules could be derived from.[15] As in the case of arithmetic for merchants, geometry for surveyors,

---

[13] Luca Pacioli: *Summa de arithmetica, geometria. Proportioni e proportionalita*, Venetia: Paganino de Paganini, 1494. Pacioli is sometimes referred to as the "Father of Accounting and Bookkeeping" although he did not actually invent bookkeeping systems. See, for instance, L. Murphy Smith (2013); Heinrich Schreiber: *Ayn new kunstlich Buech welches gar gewiss vnd behend lernet nach der gemainen Regel detre*, Nürnberg: Stuchs, 1518; Petrus Apianus: *Eyn newe unnd wohlgründte underweysung aller Kauffmanss Rechnung in dreyen büchern*. Ingolstadt: published by the author, 1527.

[14] Starting in 1518, Adam Ries published several computing books which were reissued over a hundred times. The last of these was *Rechenung nach der lenge/ auff den Linihen vnd Feder …* (Leipzig: Berwalt, 1550). See also Robert Recorde: *The grounde of artes, teaching the work and practice of arithmetice …* London: R. Wolfe, 1543; Christoph Rudolff: *Behend und hübsch Rechnung durch die kunstreichen regeln Algebre, so gemeinicklich die Coß genennt werden …* Straßburg: Cephaleus, 1525; Michael Stiffel: *Arithmetica integra*, Nürnberg: Johann Petreium, 1544. For the well-known priority dispute between Tartaglia and Hieronimo Cardano on the solution of cubic equations, see, for instance, Feldmann (1961).

[15] To name just some of these treatises from the sixteenth and early seventeenth centuries: Köbel (1522) (see note 5 above); Elie Vinet: *L'Arpanterie. Livre de geometrie, enseignant à mezurer les*

too, was a standard feature of early modern books on practical mathematics, particularly books that tried to cover the entire field of geometry. The same holds for practical mathematics for engineers, gunners, and navigators. Some of these books also dealt with mathematics for not only one but several of these practical fields.[16]

A special topic of practical geometry worthy of a brief mention is the calculation of the volumes of odd solid figures such as barrels and similar vessels. Mundane though this task was, it turned out to be difficult in practice and provoked first-rate mathematicians like the astronomer Johannes Kepler to publish writings on it.[17]

The pressing need to enable practitioners without mathematical training to accomplish tasks like surveying or bookkeeping led to the introduction, invention, and development of further aids besides books. In order to facilitate calculations, several types of *tables* were compiled and published (in books or on their own): multiplication and division tables, tables of squares, square roots and other surds. Trigonometric and logarithmic tables, first published at the beginning of the sixteenth and seventeenth centuries respectively, were an important tool, not for simple surveying tasks or bookkeeping, but in higher geodesy and astronomy.[18] The invention of *mechanical calculation instruments* was another answer to this precarious situation – instruments such as computing rods (*Rechenstäbe*) and the sector or military compass (*Proportionalzirkel*).

The computing rods or bones device, a descendant of lattice multiplication used in many different cultures, consists of rods with a square profile on which multiplication tables are

---

champs, et pluzieurs autres chozes, Bordeaux: Simon Millanges, 1577; Grosgebawer (1596) (note 5 above); Aaron Rathborne: *The surveyor in foure books*, London: W. Stansby, 1616. Erasmus Reinhold's *Gründlicher und warer Bericht vom Feldmessen* (1574) is notable for a peculiarity: The last of the book's five parts is dedicated to proving – for the reader's pleasure and diversion ("Dir zur Lust und Kurzweil") – the correctness of the procedures and rules described in the preceding parts. However, the proofs given are situational, not general since, as Reinhold added, the theoretical principles of geometry would be too demanding ("würde zu schwer sein") for the reader.

[16] To mention just some examples from the sixteenth and early seventeenth centuries: Petrus Apianus: *Instrument Buch*, Ingolstadt: by the author, 1533 book 3; Benjamin Bramer: *Trigonometria planorum mechnica*, Marburg: Egenholff, 1617, Appendix; Daniel Schwenter: *Geometria practica nova*, Nuremberg: Halbmayer, 1618–25. Leonard Digges' *Pantometria* (posthumously published by his son in 1571) dealt with practical geometry for navigators and gunners as well as surveyors. Geometry in the context of fortification was taught in Claude Flamand's *Les mathématiques et géométrie* … (Montbéliard: Foillet, 1611) and Johann Faulhaber's *Ingenieurs-Schul* (Frankfurt: Kieser, 1630–33).

[17] Johannes Kepler: *Stereometria doliorum vinariorum*, Linz: Blanck, 1615; a German version appeared in 1616. Calculation of barrel volumes also occupied other mathematicians of the time, e.g. Ulrich Kern: *Eyn new kunstliches wolgegründets Visierbuch*, Straßburg: Schäffler, 1531 or Jakob Köbel: *Rechnen und Visieren* Frankfurt: Egenolff, 1532. For literature on calculating volumes before the sixteenth century, see Peters (2018).

[18] Trigonometric tables were calculated and used in Antiquity as well as in the Arabic Middle Ages. In the West, they sprang up in the wake of *De triangulis omnimodis* by Regiomontanus (Johannes Müller 1436–1476), posthumously published in 1533. The first logarithmic tables were calculated and published by John Napier (1550–1617) (*Mirifici logarithmorum canonis descriptio* 1614) and Jost Bürgi (1552–1632) (*Arithmetische und geometrische Progresstabuln* 1620).

inscribed in such a way that multiplication and other arithmetic operations can be performed by juxtaposing them. Best known and widespread were the rods designed by John Napier and Edmund Gunter (1581–1626) at the beginning of the seventeenth century.[19] A much wider scope of applications was offered by the sector or military compass, an instrument consisting of two rulers of equal length joined by a hinge. A number of scales are inscribed upon the rulers. Based on the classic mathematical theory of proportions as well as the geometric theory of intersecting lines, these scales facilitate various mathematical calculations. It was used for solving problems in gunnery, surveying, and navigation and became the most important computing device of the seventeenth and eighteenth centuries.[20]

The theme of this book is the interrelations between the technological and scientific literature of the early modern period. In the case of literature on mathematics for practitioners, these relations seem to have been one-sided: The mathematics that this literature offered to practitioners drew explicitly, or more often, implicitly on learned mathematics either as it was taught at universities or as codified in extant classical texts such as Euclid's *Elements*. Learned mathematics, for its part, almost never received impulses for its development from this mathematics for practitioners – if it took notice of the latter at all.

However, this general picture of the relationship of learned and practical mathematics needs some qualification. As regards practical geometry derived from Euclid and adapted to the needs of practitioners, the *theorems* of Euclid's *Elements,* as already indicated, played only a minor role compared to its *problems* which presented a rich reservoir of procedures for constructing, inscribing, circumscribing etc. non-trivial geometrical figures. But as Euclid presented only a selection of such procedures known and used by architects and other practitioners in classical Greece, one has to reckon with an ongoing undercurrent of practical geometry alongside and detached from learned geometry.[21] It is therefore very likely that some of the procedures taught in the early modern mathematical literature for practitioners originated not in a learned context but among practitioners.

Perhaps the most prominent examples of a literature of this kind from around 1500 are some books (and manuscripts) on practical geometry composed not by educated authors but by master craftsmen. Architects like Matthäus Roritzer (ca. 1435–1495) or Philibert De L'Orme (ca. 1510–1570) published on geometrical procedures in connection with stereotomy and in doing so, opened a field of non-Euclidean geometry that was to become the mathematical discipline of *descriptive*

[19] For the computing rods, see John Napier: *Rabdologiae* (1617) and Edmund Gunter: *The description and use of sector, the cross-staffe, and other instruments* (1624). For lattice multiplication, see Chabert (1999).

[20] Concerning the invention of the sector, a famous priority dispute was fought out between Galileo and the Milan physician Baldassare Capra (1580–1626) – for details, see Galileo's *Operazioni del compass gheometrico e militare* (1606), and Capra's *Usus et fabrica circini cuiusdam proportionis* (1607). In view of the rich further development of the sector in the course of the seventeenth century, it makes little sense to explore this dispute: the two men could not anticipate what they had triggered and are thus merely two authors of the sector. For the sector and its development, see, for instance, Schneider (1970a), Dreier (1979) 41ff., and Damerow and Lefèvre (1985) 301–312.

[21] For practical geometry besides learned geometry in ancient Greece, see, for instance, Haselberger (1999); for such an undercurrent of practical geometry, see, for instance, Høyrup (2009).

geometry; and painters like Piero della Francesca (1416–1492) or Albrecht Dürer (1471–1528) wrote or published on geometrical procedures in connection with rendering in perspective. In doing so they opened another field of non-Euclidean geometry that was to become the mathematical discipline of *projective geometry* (see Sects. 2.5 and 2.6 of the chapter on architecture in the present volume). These examples are noteworthy because they show that the relationship of learned and practical mathematics in the early modern period was not always a one-sided matter.

7.4 Surveying Without Angular Measurement

Following the example of the Roman *Agrimensores*, early modern practical surveyors preferred surveying without angular measurement. The simplest method, applied from the late Middle Ages up to the eighteenth century, consisted in measuring distances – e.g. the edges of an acre – by means of yardsticks (the *vigra* of the ancient surveyors), a rope or a chain, and furthermore in raising perpendiculars by means of a surveying cross, the *groma* of the ancient surveyors (Fig. 7.1) Whereas the Roman surveyors used yardsticks divided according to Roman standard units of length their early modern successors had to cope with a huge variety of local units. The calibration of yardsticks, ropes or chains remained a critical issue that occupied practical mathematicians up to the seventeenth century.[22]

> Since direct measuring of distances (above all of baselines in the case of triangulations) also remained a critical issue in higher geodesy, we find many authors who discussed or proposed precise measuring devices for direct measuring of distances and their calibration: in connection with measurement of meridian sections, for example, Snellius (Willebrord van Roijen Snell 1580–1626) and Jean Picard (1620–1682) or in connection with measuring chains and their calibration, for example, Christopher Clavius or Edmund Gunter.[23]

The repugnance for angular measurement by surveyors was not due to a lack of a handy device for this purpose. The simple *geometric quadrant* – that is, a quarter circle of the old astrolabe with its limb divided into ninety degrees and furnished with sights – was known and used during the Middle Ages. As illuminated manuscripts of the fifteenth century show, it was employed particularly in siege craft and for measuring heights or depth (e.g. the heights of towers, hills, etc.).[24] Whereas the measuring of angles was an age-old standard practice of astronomers and became such a practice of navigators with the rise of oceanic navigation, the early modern development of instruments for surveying was geared to effective devices that simply allowed measuring of angles to be avoided.

---

[22] For ancient and early modern yardsticks and the surveying cross, see Schmidt (1935), sections II and III.

[23] W. Snellius: *Eratosthenes Batavus* (Leiden 1617); Picard: *Mésure de la Terre* (Paris 1671); Ch. Clavius: *Geometria practica*, book 1 (Rome 1604); E. Gunter: *Description* (see note 19 above).

[24] See, for instance, DMD – IDs tit01, tit31r, tit32 or fs72v/73r.

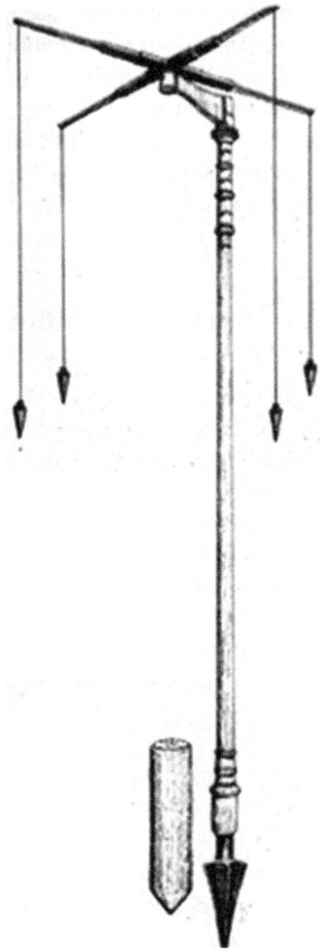

Fig. 7.1 The Roman *Groma*. (Reconstruction after a find in Pompeii by an unknown author)

One such valued and widespread device was the *geometrical quadrat* (not to be mistaken for the *geometric quadrant*) recommended and described by the renowned astronomer Georg Peuerbach (1423–1461) and by several sixteenth-century authors. Although it could also be used for measuring angles (but only in conjunction with trigonometric tables), its attraction for surveyors was that it allowed configuration simply by sighting with its index (alidade) of the requisite right-angled triangle for applying the usual rules of calculating distances or areas. In the course of the sixteenth century several other instruments of this type (with or without the use of angle measurement) were invented, even for triangulating without measuring angles, e.g. the *Triquetum* as described by Philippe Danfrie (1532–1606), Simon Stevin (1548–1620), and Leonhard Zubler (1563–1611).[25]

[25] For the "geometrical quadrat," see Schmidt (1935) 241ff. and Dreier (1979) 33f; Georg Peuerbach: *Quadratum geometricum* (posthumously Nürnberg: Stüchs, 1516); Jacobus Malconetus described the *geometrical quadrat* as late as 1700: *Selbstlehrende Geometrie* (Frankfurt: A. and P.W. Stock); for other authors, see Schmidt (1935) 247. A detailed sixteenth-century account of

The same holds for the *plane table* for topographic mapping that was invented in the sixteenth century and further developed in the following centuries. It enabled users to draw a scaled plan of estates or lands just by sighting without measuring angles. However, the calculation of areas on the basis of these plans remained a challenge for surveyors up to the eighteenth century.[26] Roughly at the same time as the plane table the *theodolite* appeared on the scene of mathematical instruments. It was to become the central device of higher geodesy. Before following up the literature about this field of practical mathematics, we will first look at literature about the mathematical practices of early modern astronomers and navigators, starting in each case with the issue of angular measurement.

7.5 Angular Measurement I – Astronomy

Measuring angles was, as already stated, a standard practice of ancient astronomers, who left two types of goniometers to their medieval and early modern successors: the *cross staff* and the *astrolabe.* They were starting points for further developments in the fifteenth and sixteenth centuries. The cross staff (also called *Jacob's staff,* from the Roman *baculus Jacobi*) implicitly took advantage of the theory of intersecting lines. We will return to it in the context of navigation. As to the astrolabe, various devices for measuring angles were derived from it, namely all the devices that involve a circular disc (full circle, semicircle, quarter circle, and other partial circles) with a rim gradated in angular degrees.[27]

Full-circle devices were in use mainly for measuring azimuth (horizontal) angles and semicircle devices for measuring altazimuth (vertical) angles; both were combined in

how to measure distances with the *geometrical quadrat* without measuring angles can be found in book III of Tartaglia's *Nova Scientia* (Venetia: Stephano da Sabio, 1537). For early modern instruments allowing triangulation without trigonometry, see Schmidt (1935) 369ff., Dreier (1979) 38ff., and the section "Exotic Instruments" in Bennett (1987) 44ff. Philippe Danfrie: *Declaration de l'usage du graphomêtre* (Paris 1597); Simon Stevin: *Practique de Geomtrie*, book II (in *Les oeuvres mathematiques de Simon Stevin*, Leiden 1634); Leonhard Zubler: *Novum Instrumentum Geometricum* (Basel 1607).

[26] The invention of the plane table is usually attributed to the mathematician and astronomer Johannes Richter (Praetorius 1537–1616). For early descriptions of the table, see, for instance, Cyprian Lucar: *A treatise named Lucar Solace* (London 1590) or Daniel Schwenter: *Mensula Praetoriana* (Nuremberg 1623). For the resistance of mathematicians to the plane table, see, for instance, Bennett (1987) 46f. Regarding the challenge of calculating areas on the basis of these plans, a curious idea is worth mentioning: The surveyor Vincent Wing (1619–1668) proposed cutting the drawn areas that were to be calculated out of the plan and comparing the weight of these pieces of paper with the known weight of a scaled standard area of paper (Richeson (1966) 123).

[27] For early modern astronomical instruments, see Repsold (1908), Zinner (1967), Dreier (1979), Bennett (1987), and Chapman (1995). For astrolabes of the medieval West, see King (2011) and Dreier (1979) 21ff.; for the transformation of the astrolabe into several devices for measuring angles in the course of the sixteenth century, see, for instance, Schmidt (1935) 273ff.; see also Kern (2010).

theodolites. Of the part-circle devices, the quadrant proved to be the most versatile. It could be equipped with an alidade or index hand, a plummet, or with sights on one of its straight rims or on its alidade. It could also be attached to other instruments such as the *geometric quadrat* or the *Triquetum* mentioned above. Rulers could be attached to its straight rims and extended for insertion into cannon muzzles – thus becoming the *gunner's quadrant.* A rotatable straight edge with sights and plummet could be mounted on its apex to form a *mason's level* or *Setzwaage*. Semicircle discs combined with a ruler at the base and/or integrated into a rectangular triangle are in everyday use up to the present – our protractor.

Of these instruments, we will focus on the development of the quadrant as the most important astronomical instrument in the early modern period, that is, on the metamorphosis of the humble fifteenth-century instrument into the "high tech" mural quadrants of the eighteenth century like the ones by George Graham (1673–1751) and John Bird (1709–1776) for the Royal Observatory in Greenwich. This development was spurred by attempts to overcome the lack of accuracy of astronomical observations, which produced observational data that was unacceptable in the long run for both theoretical and practical reasons.

As is well known, in the medieval West astronomy was of *practical* significance for the calculation of the calendars and in particular for the calculation of the Easter date for a particular year. That is probably the main reason why astronomy was practiced continuously – though with changing intensity and quality – throughout the Middle Ages. The ephemerides calculated for this and other practical purposes, e.g. casting horoscopes,[28] depended on astronomical tables like the Alfonsine tables of the thirteenth century which increasingly deviated over time from actual observations due to an inaccurate underlying determination of the length of the sidereal year. Improved astronomical tables were urgently needed and actually calculated in the mid-sixteenth century and again at the beginning of the seventeenth century. The replacement of the Roman (Julian) calendar by the Gregorian at the end of the sixteenth century was the crowning and most far-reaching result of the efforts of practical astronomers like Christopher Clavius.[29]

As regards *theoretical,* scientific astronomy, the consistency between observations and theory was a crucial problem since its very beginnings in ancient Greece. The famous Platonic call to "save the phenomena" reflected nothing else than the intolerable fact that the observed motion of the planets (including the moon and sun) did not conform with the motion required by cosmological theory, that is, it deviated from a uniform motion on exact circular orbits. Ptolemy's model with its *deferents* and *epicycles* was not seen as a reconciliation between observational astronomy and classic cosmology but rather as the establishment of a mathematical astronomy alongside the true cosmological astronomy. Discontent with this situation was one of the main reasons that motivated Nicolaus Copernicus (1473–1543) to try another cosmological model in which the observed orbits

---

[28] For astrology in the Middle Ages, see, for instance, White (1978) 331ff.

[29] The *Prutenic tables* were published by Erasmus Reinhold (1511–1553) the Elder in 1551, and the *Rudolphine tables* by Kepler in 1627. The beginnings of the Gregorian calendar reform can be traced back to a proposal by Aloysius Lilius (ca. 1510–1576) which was elaborated mainly by Clavius; see Ch. Clavius: *Novi calendarii romani apologia* (Rome 1588) and *Romani calendarii a Gregorio XIII P.M. restituti explicatio* (Rome 1603). See Coyne et al. (eds.) (1983).

corresponded more closely to the ideal of circular motions – by taking the sun as their center.[30]

Against this background it becomes understandable why the endeavors toward improved observations and greater accuracy became a major concern of early modern astronomers, fueling their demand for better instruments and efforts to design improved instruments.

Until well into the seventeenth century, astronomical and other mathematical instruments were designed or, in the case of existing instruments, further developed by scientists or practical mathematicians, not by the carpenters, locksmiths, and engravers who manufactured them. Accordingly, the literature on astronomical instruments was composed and/or published by astronomers like Peuerbach, Regiomontanus, Apianus or Tycho Brahe (1546–1601). Only later do we meet with instrument makers who outgrew their traditional subservient role and took the initiative in developing and improving mathematical instruments. Three early modern books that provide an overview of the entire range of mathematical instruments were written by such advanced craftsmen, the instrument makers Nicolas Bion (1652–1733), Jacob Leupold (1674–1727), and George Adams Jr. (1750–1795).[31]

Tycho Brahe[32] is certainly the most prominent of the early modern astronomers involved in the improvement of instruments. He is famous for positional observations that set a new standard of accuracy, and in particular for his observations of the orbit of the planet Mars. As is generally known, this constituted the empirical basis on which Kepler established his theory of elliptic planetary orbits and, ultimately, his famous three laws. Tycho Brahe is equally and deservedly famous as the designer of improved astronomical instruments that facilitated his achievements in positional observations. He published information on his improvements of astronomical instruments in two treatises: *Astronomiae instauratae mechanica* (1598) (Fig. 7.2) and *Astronomiae instauratae progymnasmata* (1602).

The crucial problem involved in improving the astronomical instruments of the time was to find new ways of subdividing the degrees engraved on their limbs. One (not so new) way was to enlarge the instruments and thus the space between the degree marks. The other way was the introduction of new kinds of engraved scales or of additional sliding scales – transversal or diagonal scales and/or Vernier scales

---

[30] It was due to this tradition of two coexisting astronomies – a true cosmological one and a merely mathematical one – that, in the first couple of decades after the publication of Copernicus' *De revolutionibus orbium coelestium* in 1543, the heliocentric system was tolerated by the Roman Catholic Church as just another mathematical hypothesis. Its ban began when it was taken as true cosmology, as by Giordano Bruno (1548–1600) or later by Galileo. For the so-called Copernican Revolution, see, e.g., Renn (2020) 120ff.

[31] For a survey of the development of the trade of instrument makers in the early modern period, see Price (1957), Bennet (1987) chaps. 5 and 6., Schneider (1970b) 237, 242, and Damerow and Lefèvre (1985) 10ff. Nicolas Bion: *Traité de la construction et des principaux usages des instrumens de mathematique* (Paris: Chez La Veuve de J. Boudot, J. Collombat, et J. Boudot fils, 1709); Jacob Leupold: *Theatrum aithmetico-gometricum* (Leipzig: Christoph Zunkel, 1727); George Adams: *Geometrical and graphical essays* (London: published for the author by R. Hindsmarsh, 1791).

[32] For the following, see Dreyer (1972), Bennett (1987) 23ff., and Chapman (1995) chap. 2.

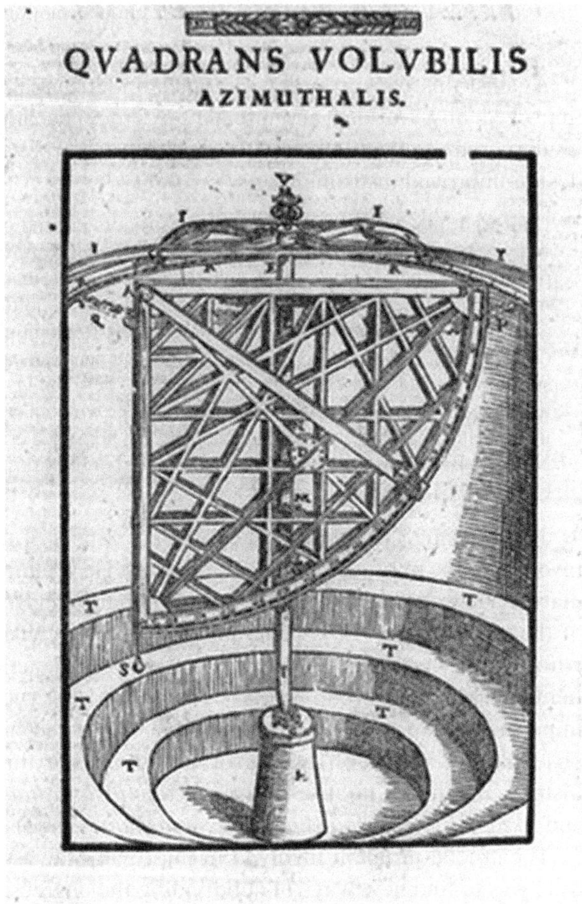

Fig. 7.2 Tycho Brahe's Azimuth Quadrant (Brahe 1972, p. 23)

(*Nonius*) – that allowed fractions (minutes, or even seconds) of degrees to be read off.

Huge quadrants had already been built and used in the Middle Ages, e.g. by the Timurid sultan and astronomer Ulugh Beg (1394–1449) in Samarkand. Tycho himself designed a mural quadrant in Uraniborg with a radius of about two meters.[33] These big instruments were further developed in the course of the seventeenth and eighteenth century. Their frame could be suspended rotatably at their apex – for instance, by Johannes Hevelius (1611–1687)[34] – or fixed to a wall as in the observatories of the eighteenth century.

Linear transversals drawn on rulers or other straight parts of instruments as a means of reading off fractions of units were known and used at the end of the Middle Ages. The application of circular transversals drawn on arcs seems to have been a sixteenth-century development that Tycho took up. After him it became a standard scale on astronomical

[33] Tycho had earlier built a quadrant in Augsburg that was more than twice as big; see Thoren (1973) 25.

[34] Hevelius described his instruments in *Machina coelestis* (part I, Danzig: Reiniger 1673).

instruments up to the eighteenth century. The Vernier scale (*Nonius*) can be traced back to a proposal of the Portuguese mathematician Pedro Nuñez (1502–1578) that was not very practical but was developed further by several authors before the French mathematician Pierre Vernier (1580–1637) established its classical shape in 1631.[35]

The implementation of these new kinds of engravings on the limbs of quadrants and other circular instruments since the end of the sixteenth century led to an increasing accuracy that amounted to a watershed in the history of astronomical instruments. By comparison, equipping these instruments with telescopic sights in the course of seventeenth century was certainly a less momentous breakthrough.[36] But the use of telescopic sights, together with the new level of accuracy, exposed the instruments' many remaining weak points or problematic aspects – the lack of smoothness of their mechanics, the heavy weight of the big instruments, which made their frame prone to deformations, the impact of changing temperatures and degrees of humidity on the metal limbs, and so on.

The problem of parallax-free sights became more important with the introduction of telescopic sights and led to the invention of the reticule. The invention of the micrometer screw helped to solve other sighting problems, for instance in connection with measuring the sun's diameter by projections with the camera obscura. The attempt by Robert Hooke (1635–1703) to replace divided rims by toothed ones driven by worms and a micrometer proved impractical.[37]

The engraving of precise dividing marks remained a serious challenge well into the eighteenth century or, more precisely, as long as it was done by hand. There were no unsolved mathematical problems in connection with dividing arcs although several tricky mathematical strategies were contrived and used for this purpose. The point of impasse was the material action of precisely and consistently marking and engraving lines, dots and arcs by means of incising tools, several kinds of compasses and so on. The pre-industrial fabrication of mathematical instruments had reached its limits. They were only transcended with the arrival of dividing machines such as those invented by Jesse Ramsden (1735–1800) and Georg Friedrich Brander (1713–1783) in the last third of the eighteenth century.[38]

---

[35] See, for instance, Zinner (1931) 574f. Pedro Nuñez: *De Crepusculis*. Olyssippone: Ludovicus Rodericus, 1542; Pierre Vernier: *La construction, l'usage, et les propriétés du quadrant nouveau de mathématique*. Brussels: F. Vivien, 1631.

[36] After Galileo's use of his telescope for astronomical observations in 1610, it took time before astronomical instruments were routinely furnished with telescopic sights. As late as in the 1670s and 1680s, Hevelius continued to perform observations without a telescopic sight and obtained observational data as accurate as that of other contemporary astronomers equipped with sights; see Saridakis (2001) chap. 4; see also Chapman (1995) 38ff.

[37] The reticule, the invention of which is usually associated with the name of the astronomer and mathematician William Gascoigne (1612–1644), became a standard feature of sighting telescopes only at the end of the seventeenth century. For Gascoigne, see, for example, Bennett (1987) 63f. For Robert Hooke's instrument with a toothed rim, see his *Animadversions on the first part of the* Machina coelestis *of … Johannes Hevelius* (London 1674); see also Chapman (1965) 45ff.

[38] See, for instance, Repsold (1908) 79ff and Brachner (1983) 349.

> In 1767 the instrument maker John Bird, who built the new mural quadrant for the observatory at Greenwich in 1750, published a treatise on his methods and strategy of dividing – "by order of the Commissioners of Longitude." He was obliged, furthermore, to commit himself to instruct an apprentice in the art of dividing for seven years. This provision shows not only the Commissioners' mistrust regarding the completeness and openness of his treatise but also their insight that the art of dividing involved skills and practical experiences that no treatise can transmit.[39]

With regard to the general theme of this book, we have to ask which interrelations between the technological and scientific literature of the early modern period we observed in the case of early modern astronomy. The early modern technological literature on calendar making, ephemerides, and astronomical tables, as well as on instruments and methods of astronomical observation, owed almost nothing to the contemporary scientific astronomical literature, that is, to the cosmological literature in the Platonic/Aristotelian tradition. The topics of this technological literature were neutral with respect to a geocentric or heliocentric cosmos. The cosmology, on the other hand, proved not to be neutral with respect to the observations made by early modern practical astronomers. Rather, on the basis of the observational data obtained since the mid-sixteenth century, a new theoretical astronomy emerged that questioned and eventually displaced the old cosmology.[40] This new scientific astronomy benefited from the advancement of observational practices and gradually affected the agenda (objects) of astronomical observations, without, however, being beneficial for these practices. This relationship started to become more mutual in the eighteenth century when the most difficult astronomical tables became calculable on the basis of Newton's astronomy, including the moon tables needed by navigators for the determination of longitudinal positions at sea.

7.6 Angular Measurement II – Navigation and Mathematical Geography

Up to the end of the fifteenth century, the art of navigation was not a field of practical mathematics. It became so with the commencement of oceanic seafaring, which made use of techniques as well as instruments of practical astronomy for coping with spatial and temporal orientation (localizing position and determining time) as well as finding and keeping appropriate courses far away from coasts.

[39] John Bird: *The method of dividing astronomical instruments*. London: John Nourse, 1767. For the strategy of dividing used by Bird, see, for example, Damerow and Lefèvre (1985) 321.

[40] Kepler's analysis of Tycho's observational data questioned not just a detail of traditional cosmology – the dogma of the planets' uniform motion on exact circular orbits. He actually questioned this cosmology's essential core, that is, its ontological separation of a translunar sphere from our sublunary world, and opened the way for an astronomical science that tried to understand celestial phenomena by the same natural laws as phenomena on Earth. This was the path from his own *Astronomia nova* (1609) via Newton's *Principia* (1687) to modern astrophysics.

Coastal seafaring in the Middle Ages was not in need of such mathematical resources but could make do with simpler aids. The instruments used by the seamen of the period comprised the *sandglass*, *magnetic compass* (since the thirteenth century), and *sounding log*. They also used mariners' handbooks – the *Rutter* (*Roteiro*, *Routier*) – informing them in words and pictures about peculiarities of harbors, bays, and capes, as well as about soundings, drifts, tides, etc. of these places. And they used a particular kind of sea chart, called *portolan charts*, maps without a geographical coordinate system.[41]

Latitude sailing was the favored practice at the beginning of oceanic navigation. A ship's latitudinal position could be easily checked by measuring the altitude of the sun at noon and the polar star at midnight. For this sailing method, a navigator needed (1) a basic familiarity with astronomical methods of localization as well as with the principles of mathematical geography; he also needed (2) suitable instruments for angular measurements; and (3) a couple of tables and manuals.

(1) The question of procuring trained oceanic navigators led to the establishment of a special school as early as 1503 – the *Casa de Contratación* in Sevilla. Experienced mariners aspiring to become navigators could also find instructions in the emerging textbook literature on navigation.[42]

(2) Up to the eighteenth century, mariners preferred the cross staff (Jacob's staff) as the appropriate instrument for measuring angles at sea. Originating from the use of the cross staff for measuring the sun's altitude by back observation, that is, by projecting the shadow of the instrument's crosspiece onto the gradated main staff, the instrument was further developed into the back staff or Davis quadrant. Eventually these were replaced by reflecting sextants and octants devised by several English astronomers at the turn of the seventeenth century.[43]

---

[41] Medieval mariners' manuals succeeded these manuals of ancient navigators, of which several Greek manuscripts with the title *Periplus* are extant. The oldest medieval manual is the *Compasso da Navigare* (Codice Hamilton 396) from the thirteenth century; an example of fifteenth-century manuals known for its remarkable views of coastal strips is *Grand Routtier* by Pierre Carcie (1441–1502); see Dreier (1979) 51f. Below in this section we will briefly address the significance of mariners' manuals for early modern cartography. For medieval portolan charts, see, for instance, Kretschmer (1909). Mariners' manuals and portolan charts were both used by coastal mariners well into the seventeenth century. For the development of the magnetic compass and the compass rose in the early modern period, see, for instance, Dreier (1979) 58 and Bennet (1987) 27ff.

[42] For the training at the school in Sevilla, see, for instance, Collins (2014). The early modern textbook literature for mariners emerged in fifteenth and sixteenth-century Portugal and Spain, then the leading kingdoms with overseas colonies. The most influential of these textbooks were Pedro de Medina's (1493–1567) *Arte de navegar* (Valladolid 1545) and Martin Cortés de Albacar's (1510–1582) *Breve compendio de la sphera y de la arte de navegar* (Sevilla 1551). These Spanish books were translated and expanded in the Netherlands, France, and England in the course of the sixteenth and seventeenth centuries. To name just some of the textbooks in the wake of Medina and Cortés: Nicolas de Nicolai (1517–1583, trans. and ed.): *L'art de naviguer de maistre Pierre de* Medina (Lyon, 1553); William Bourne (c. 1535–1582): *A regiment for the Sea* (London 1574); Lukas Janszoon Waghenaer (1533–1606): *Spiegel der Zeevaerdt* (Leyden 1584); John Davis (1550–1605): *The Seaman's Secrets* (London 1595).

[43] For the cross staff, see Schmidt (1935) 338ff., Dreier (1979) 64ff., Bennett (1987) 34ff., Chapman (1995) 24ff. The instrument's straight main staff was marked by projecting the degrees of a divided circle onto it; for an early description of this method, see Petrus Apianus: *Cosmographicus liber*

(3) The equipment of oceanic navigators included, besides ephemerides, sun tables for reading off the latitude that corresponds the measured altitude of the sun on a particular day, sea charts or tables from which the latitude of the destination port could be taken, and tables with the longitude distances on a particular degree of latitude (*Breitenmaßstab*).[44]

As such tables of longitude distances show, questions pertaining to mathematical geography were already important for latitude sailing and much more so for navigating on courses of any desired bearing. As is well known, mathematical geography – the science of methods of presenting the Earth on maps by projecting the terrestrial globe onto a plane, as well as of various resulting coordinate systems – can be traced back to Hellenistic Antiquity.[45] It became known to the West by the rediscovery of Ptolemy's *Geographia* in the fifteenth century. The projections of the terrestrial globe onto a cylinder jacket that Ptolemy proposed in his *Geographia* yielded coordinate systems which represented distances fairly proportionately (*längentreu*). However, these systems of projected geographic coordinates turned out to be unsatisfactory for the world maps and sea charts created as a result of Christopher Columbus' crossing of the Atlantic Ocean and Vasco da Gama's discovery of the oceanic route from Europe to India in the 1490s.[46] Thus, in sixteenth century, new systems of projected geographical coordinates were proposed, including a highly significant one by the Dutch cartographer Gerhard Mercator (1512–1594) that took into account the special needs of oceanic navigation.[47] The

(Landshut 1524). John Davis invented the instrument known as the Davis quadrant in the 1590s. Among the inventors of reflecting instruments, the best known are Isaac Newton, Edmond Halley (1656–1742) and John Hadley (1682–1744).

[44] In late fifteenth and early sixteenth century, most of these resources were composed by Portuguese mathematicians; also important was Regiomontanus' *Calendarium* (Nuremberg 1474, Venice 1476, and further editions in several vernaculars) which contained ephemerides and other highly accurate tables.

[45] For a history of geographic projection methods from Ptolemy up to the twentieth century, see Snyder (1993).

[46] For the productive reception and appropriation of Ptolemy's coordinate systems by cartographers at the turn of the fifteenth century, see Bennett (1998) 204ff. We should note that from the end of the fifteenth century mariners used another kind of sea chart besides portolan charts, namely *Plattkarten*, that is, charts based on an equirectangular projection which were recommended by Ptolemy; see Schneider (1970b) 234 and Dreier (1979) 52f.; for equirectangular projections and the three projections proposed by Ptolemy, see also Snyder (1993) 5f. and 10ff.

[47] Gerhard Mercator: *Nova et aucta orbis terrae descriptio ad usum navigantium*. (Basel 1569). For Mercator's map, see Krücken (1994). For an early demonstration of the mathematical basis of Mercator's projection, see *Certain Errors in Navigation* (London 1599) by Edward Wright (1561–1615). For the Mercator projection and for Wright's discussion of it, see Snyder (1993) 43ff. and 47ff. To name just a few other famous world maps published in the first half of the sixteenth century: Martin Waldseemüller's (ca. 1472–1520) *Universalis cosmographia secundum Ptholomaei traditionem et Americi Vespucii aliorumque lustrationes* (map, globe and treatise) 1507; Johannes Werner's (Vernerus 1468–1522) rendering of the globe in the heart-shaped Stab-Werner projection in his *Nova translation [...] Libellus de quattuor terrarium orbis in plano figurationibus* 1514; Oronce Finé's (1494–1555) *Nova universi orbis descriptio* (1531; his first

projection proposed and epitomized by Mercator's world map of 1569 was a conformal (*winkeltreue*) projection that allowed any course of constant bearing on a chart to be represented as a straight line.

The problem of nautical concern that Mercator's projection solved was the representation of loxodromes on sea charts as straight lines. Loxodromes are the curves a ship describes when sailing a course of constant bearing which intersects all meridians at the same angle (right angle excluded). These curves do not describe a great circle but a spiral, as the Portuguese mathematician Pedro Nuñez (1502–1576) discovered and demonstrated mathematically.[48]

The[49] map projections of the sixteenth and seventeenth centuries could be, and actually were realized by geometrical constructions which were supported by calculations with the aid of logarithm tables from the beginning of the seventeenth century. These mathematical means proved to be insufficient in the light of further developments in mathematical geography, particularly since it had become clear that the terrestrial globe was not a sphere but an ellipsoid. This spurred eighteenth-century mathematicians like Leonhard Euler (1707–1783), Johann Heinrich Lambert (1728–1777), and Joseph-Louis Lagrange (1736–1813), using the newly invented calculus, to find formulas for calculating such projections.[50]

We will not explore the art of designing, drawing, printing and fabricating maps and globes that emerged and flourished in the sixteenth century.[51] Instead we will confine ourselves to pointing out the remarkable interplay that can be seen here between two branches of practical mathematics, navigation and cartography. Cartographers provided seamen with sea charts that facilitated navigation, and seamen provided cartographers with the topographical data of the new regions they detected that enabled the design of world maps and globes. The world maps and globes of the sixteenth century show, however, that the underlying topographical data left much to be desired, as could be expected in view of the many contingent circumstances in the early phase of the Western age of geographical discoveries.

Seamen[52] sailing to previously unknown regions routinely produced maps and views of the coasts and ground plans of the harbors they encountered. In other words, they supplied the material for mariners' handbooks. In contrast to land surveyors, surveying seamen could

---

world map appeared in 1519). – The first-ever world map, with only rudimentary coordinates, was published by Juan de la Cosa in 1500. Juan was the helmsman of the *Santa Maria* on which Columbus crossed the Atlantic Ocean for the first time in 1492. For sixteenth-century Dutch sea charts, see, for instance, Dreier (1979) 52f.

[48] Pedro Nuñez: *Tratado em defensam da carta de marear*. Lisboa 1537.

[49] For the following, see Snyder (1993) 62ff.

[50] Leonhard Euler: *De projectione geographica superficiei sphaericae,* in: Actademiae Scientiarum Imperialis Petripolitanae 1777/78; Johann Heinrich Lambert: *Anmerkungen und Zusätze zur Entwerfung der Land- und Himmelscharten* (1772), ed. by Albert Wangerin (Leipzig: Engelmann, 1894); Joseph-Louis Lagrange: *Sur la construction des cartes géographiques*, in: *Nouveaux mémoirs de l'Academie Royale des Sciences et Belles-lettres* (1779) 161–210. For Lambert's projections, see Snyder (1993) 76ff.

[51] For early modern cartography, see, for instance, Buisseret (2003) and Bagrow (2017). For early modern maritime cartography, see, for instance, Whitfield (1996) and Schilder (2017).

[52] For the following, see Damerow and Lefèvre (1985) 351–61.

not avoid angular measurements. Since in most cases direct measurement of distances was impossible, the data for the maps and ground plans of coasts and harbors could only be obtained by some kind of triangulation. But angular measurements were a familiar task for navigators. The geographical data collected in mariners' handbooks constituted the basis of charts and maps of the newly discovered regions – maps that were often classified as secret documents by the governments of the competing colonial powers.

Aside from particular inaccuracies, however, the early modern maps and globes also show a systematic defect: Whereas the north-south extension and distances are fairly correctly represented, the maps and globes almost always appear to be considerably or even grotesquely distorted as regards east-west extensions and distances. This defect was due to the fact that no reliably practical method of determining the longitudinal position of a spot had yet been found.[53]

The longitude problem did not raise any theoretical questions for mathematical geography, according to which the equator and each circle of latitude were divided into 360 degrees intersected by meridians or lines of longitude. Rather, the longitude problem was a practical one. In contrast to determining the latitude of a spot, there was no simple measurement of a celestial altitude for finding the longitudinal position of a particular spot. Knowing that the west-east angular velocity of the Earth is 15 degrees an hour, a longitudinal position could easily be found if one knew the actual time of day at a reference place. But how could one know the time hundreds or thousands of miles away from such a reference place?

To solve this longitude problem it was necessary to find a practicable method for determining the time of day at a reference place (in an age without radio communications, satellites, etc.).[54] One method that suggested itself was trying to design and fabricate a transportable clock that ran accurately for several months or years without readjusting – a huge and almost unmanageable challenge to pre-industrial clock making.[55] Another suggested method was to use the fact that celestial constellations can be observed at the same moment at the reference place and at any place on Earth. Only one such constellation changes fast enough that a specific moment of time can be attached to it, namely the relationship of the moon to any fixed stars. The calculation of tables of these constellations also proved to be a huge and almost unmanageable challenge – this time in the field of theoretical astronomy and higher mathematics.[56]

---

[53] This problem was so urgent for seafaring nations that special boards for its solution were established at the turn of the seventeenth century. It goes without saying that the traditional method of dead reckoning (*Koppelnavigation*) which is subject to accumulative errors was not reliable. For this method and the traverse board it used, see Bennett (1987) 30, and Damerow and Lefèvre (1985) 345f.

[54] For the following, see, for instance, Schneider (1970b) 233ff. and Bennett (1987) 53ff. For the abundance of small problems that had to be solved in this context, see Damerow and Lefèvre (1985) 338–52.

[55] John Harrison (1693–1776) fabricated the first chronometer that met these requirements to an astonishing degree. For Harrison, see *The Principles of Mr. Harrison's Time-Keeper*, published by the Commissioners of Longitude (London 1767): https://archive.org/details/principlesmrhar00unkngoog/page/n6

[56] Drawing on Leonhard Euler's work on the lunar movement, Tobias Mayer (1723–1762) calculated moon tables of the required accuracy. For Euler's moon theory, see Verdun (2014)

7.7 Higher Geodesy

The remarkable interplay between navigation and cartography we addressed above can be extended to include mathematical astronomy as well as geodesy. As is well known, the principal elements or parameters of the terrestrial coordinate system are derivations or projections of the principal elements or parameters of the celestial coordinate system established in Hellenistic Antiquity – celestial/terrestrial poles, celestial/terrestrial equators, ecliptic/northern and southern tropics, and so on. According to the terrestrial coordinate system of early modern mathematical geography and the various cartographic projections based on this system, the terrestrial globe was always assumed to be an exact geometrical sphere. This assumption was bound to be seen sooner or later as a mere mathematical model by geographers and cartographers who had to take into account the actual situation on the Earth's surface. Even mathematical geographers had to resort to some kind of geodetic measurements for determining the dimensions of their ideal geometrical sphere. The measurement of a meridian segment seemed to offer the most feasible way of determining these dimensions.

This measurement suggested itself because one can determine by familiar astronomical methods, namely, measuring the altitude of the sun at noon gives the distance in arc degrees between two places on one and the same meridian. However, ensuring that both places actually lie on the same meridian as well as measuring their distance are not trivial tasks and can hardly be accomplished by the methods of practical surveyors because they demand geodetic triangulation.[57]

A particular aspect of the terrestrial coordinate system's dimensions must be mentioned because of its practical significance for cartography and navigation, namely, the length of the distance between degrees of longitude (east-west distance). Obviously, this length varies with the parallel of the respective latitude – it is greatest at the equator and becomes zero at the pole. As this length is an element of a mathematical model, its determination for a particular parallel of latitude is not an issue of measuring but of calculation. Tables with these lengths for each parallel of latitude (*Breitenmaßstab*) were calculated and published starting from the end of the fifteenth century.[58]

---

section 3.2.4; for Mayer, see Hutton (1795) II 56, Wepster (2010), and Weißbecker (2012).

[57] In order to determine the circumference of the Earth, such a meridian measurement was already performed in Hellenistic antiquity by an exemplary practical mathematician, namely by the astronomer and geographer Eratosthenes (ca. 276 BC – ca. 195 BC). For meridian measurements in the Arabic Middle Ages, see Schmidt (1935) 164f. The first early modern measurement of a meridian arc segment by triangulation was performed in 1615 by the Dutch mathematician Willebrord Snellius; he published its results in his *Eratosthenes Batavus seu de terrae ambitus vera quantitate* (Leiden 1617). See, for instance, Haasbroek (1968) 59ff. We will return to early modern meridian measurements below.

[58] Fifteenth-century Portuguese mariners' manuals already provided such tables. The tables that Martin Cortés de Albacar published in his *Breve compendio de la sphere y de la arte de navigar* (1551 – see note 42 above) were adopted by mariners' manuals in several European countries. The mathematical problems involved in the calculation of these tables remained a challenge for mathematicians from the sixteenth century, e.g. Edward Wright, Thomas Harriot (1560–1621), up to the eighteenth century, e.g. Edmond Halley. See, for instance, Dreier (1979) 54, and Schneider (1970b) 132f.

In the early modern period the demand for geodetic measurements did not arise first in connection with meridian measurements but with the development of cartography from the beginning of the sixteenth century. Surveying entire regions or countries became a necessity as soon as cartographers wanted to construct maps drawn to any suitable scale so that information about distances could be read off these maps. Surveying in this order of magnitude only yields reliable results when performed by triangulation.

The triangulation method can be traced back to ancient Greek as well as medieval Arabic sources. In the early modern period, it was the Dutch physician Gemma Frisius (1508–1555) who proposed and described it as a key technique in connection with map making in his *Libellus de locorum describendorum ratione* (1533) – a book that was instrumental for the spread of this surveying method over many European countries in the following period.[59]

The key instrument for angular measurements used in geodetic triangulations was, and still is, the theodolite. Like the method of triangulation, this instrument, too, can be traced back to forerunners in Antiquity and the Middle Ages. It was further developed in the sixteenth century and acquired its early modern standard shape – a full-circle device for measuring azimuth (horizontal) angles combined with a semicircle device for altazimuth (vertical) angles, sights, compass, and tripod. To become the "high-tech" instrument of the eighteenth century, the theodolite had to overcome similar problems regarding the instrument's mechanics, parallax-free sights, precise division of the limbs of the circle devices, etc., as discussed above in connection with the development of quadrants.[60]

In the beginning,[61] that is in the sixteenth century and the first decades of the seventeenth century, it was private entrepreneurs who performed geodetic surveys for procuring the data needed by cartographers. The cartographer Christopher Saxton (1542–1606), for example, author of the first atlas with printed maps of England and Wales, published in 1579, collected the needed data for this work through his own geodetic surveying. Such private geodetic surveying of entire regions or countries was banned for reasons of secrecy by the belligerent states of the seventeenth century. This makes it all the more astonishing that the main players of absolutist Europe only began to organize surveying of their territories for military and fiscal purposes in the eighteenth century.[62] Interestingly, the eighteenth-century topographic surveys for the Prussian and Austrian military (resulting in the

[59] Gemma Frisius'*Libellus* first appeared in Antwerp (apud Ioannes Grapheus, 1533). It became known to a broad readership when reissued as an appendix in Petrus Apianus' *Cosmographia* of 1540. Frisius, together with a goldsmith, managed a workshop for the production of globes.

[60] The *Dioptra* of Hero of Alexandria seems to be the oldest known instrument that can be rightly regarded as a kind of theodolite. For the *Dioptra*, see, for instance, Drachmann (1969). The invention of an early modern theodolite can be ascribed to Leonard Digges, namely for the *topographicall instrument* described in his *Pantometria*, published by his son Thomas in 1571 (see note 16 above). For the development of the instrument from the sixteenth century until Jesse Ramsden's famous theodolites at the end of the eighteenth century, see Dreier (1979) 34f., Bennett (1987) 145ff.

[61] For the following, see Damerow and Lefèvre (1985) 335f.

[62] Early topographic surveys of territories like that of the Electorate of Saxony (1586–1607) or of Piedmont (begun in 1622) were exceptional at the time. For state surveys of entire territories in the late seventeenth and in the eighteenth centuries, see, for instance, Torge (2007) section 5.2.

Preussische Kabinettskarte, created in 1767–1787, and the *Josephinische Mappierung*, created in 1763–1787) were performed by means of plane tables. The national surveys of England by the Scottish military engineer General William Roy (1726–1790) and of France by César François Cassini (1714–1784) were performed by triangulation with theodolites and resulted in setting up national triangle nets.

These national triangle nets became the basis for a huge geodetic enterprise of the late eighteenth century, namely the measurement of the distance between the observatories in Greenwich and Paris. This enterprise was of importance for astronomy, geography, and particularly for navigation since the nautical tables of the time referred to the meridian of either the one or the other of these observatories as the zero meridian. The exact determination of the distance between the two observatories was the essential prerequisite for translating the data in one table into those in the other.

> In this enterprise, the geodetic survey on the English side was performed by William Roy, who spent a number of months taking the direct measurement of the baseline close to Greenwich.[63] This detail is worth mentioning because it highlights the direct measuring of the baseline as one of the critical points of surveying by triangulation. The other was, of course, the precision of the theodolites used. Roy used theodolites custom-made by Jesse Ramsden. The mathematical techniques for taking into account the spherical shape of the terrestrial globe was a further critical point of large triangulations – a problem that was first tackled at the beginning of the nineteenth century.[64]

In closing this section, I would like briefly to address the famous geodetic efforts undertaken in the first half of the eighteenth century for deciding a (at that time) purely scientific issue, namely, the question of whether the terrestrial globe is flattened at the equator or the poles. Conjectures[65] that the terrestrial globe might be an ellipsoid rather than a sphere were reinforced in the seventeenth century: In 1666 the astronomer Giovanni Domenico Cassini (1625–1712) had observed that the planet Jupiter is flattened at its poles; and the astronomer Jean Richter (1630–1696) had discovered by means of a one-second pendulum during a stay in Cayenne in the 1670s that the force of gravity intensifies when approaching the equator; finally, Isaac Newton, in his *Principia* of 1687, argued that the terrestrial globe must be flattened at its poles. However, French measurements of the Paris meridian made between 1668 and 1718 suggested that the globe was flattened at the equator rather than at the poles. To resolve this question, in 1730 the French Academy of Science decided to equip and send two expeditions tasked with measuring a meridian arc close to the North Pole (in Lapland) and a meridian arc close to the equator (in Peru, today in Ecuador) – expeditions which confirmed Newton's assumption.[66]

---

[63] See Richeson (1966) 176.

[64] See, for instance, Johann Georg von Soldner (1776–1833): *Theorie der Landvermessung* (1810), ed. by J. Frischauf (Leipzig: Engelmann 1911).

[65] For the following, see, for instance, Torge (2007) section 5.1.1.

[66] Jean Picard (1620–1682) initiated the French survey of the Paris meridian, measuring the section Paris-Amiens from 1668 to 1670; see Jean Picard: *Mésure de la Terre*, Paris: l'imprimerie Royale, 1671. The surveys of the northern section, Amiens-Dunkirk, and of the southern section, Paris-

7.8 Conclusion

In concluding this chapter on practical mathematics, we should begin by recognizing that our investigations of the interrelations between the technological and scientific literature of the early modern period often reach their limits here, since technology and science became indistinguishable in several fields of practical mathematics. Among the early modern mathematical practitioners we found not only intermediaries between science and technology but also hybrid experts such as Nuñez, Mercator, Brahe, or Galileo, that is, figures who contributed to developments in technology as well as science; and we found, for example, astronomical or mathematical geographical literature that is both technological and scientific.

In the case of the mathematical tools, aids, and resources composed or invented and produced by practical mathematicians for merchants, surveyors, navigators etc., at first glance the relations between technological and scientific literature appear one-sided: The practical mathematics on offer drew explicitly or implicitly on learned mathematics, whereas the latter almost never received impulses from the mathematics for practitioners. But this is not the full picture. The development and refinement of mathematical tools like logarithm tables proved to be of significance for learned mathematics, too. Furthermore, in the context of stone masonry and perspectival depiction, a practitioners' constructive geometry existed alongside the learned Euclidean geometry and was developed into new non-Euclidean mathematical disciplines.

Finally, as regards astronomical literature, we should keep in mind that seen from the perspective of the true (Platonic/Aristotelian) cosmology, all of the efforts and achievements from Ptolemy to Copernicus were mere alterations of mathematical hypotheses and not contributions to science in the proper sense, that is, to insight into Nature. There was no relationship of mutual exchange between this cosmology and early modern astronomy. Rather, the independent development of the latter ultimately had the consequence that the classic cosmology was simply dropped and astronomy in today's sense took its place. The same holds more or less for the relationship between mathematical geography and traditional cosmography. Regarding early modern geographical literature related to astronomy and mathematics, it is usually far from clear whether a document like a table of ephemerides or a map, for example, should be classified as technological or scientific literature.

Perpignan, were carried out between 1684 and 1718 by Giovanni Domenico Cassini and his son Jacques Cassini (1677–1756). The expedition to Peru was undertaken from 1735 to 1744 by Pierre Bouguer (1698–1758), Charles Marie de Condamine (1701–1774), and Louis Godin (1704–1760); the expedition to Lapland was undertaken from 1736 to 1737 by Pierre Louis Moreau de Maupertuis (1698–1795) and Alexis Claude Clairaut (1713–1765). The scientific analysis of the expedition findings was published by Maupertuis and Bouguer: Pierre Louis Moreau de Maupertuis: *La figure de la terre déterminée par les observations de messieurs Maupertuis, Clairaut, Camus, Le Monnier et Outhier accompagnés de M. Celsius, faîtes par ordre du roy au cercle polaire,* Paris: l'imprimerie Royale, 1738; Pierre Bouguer: *La figure de la terre: déterminée par les observations de messieurs Bouguer & de la Condamine*, Paris: Charles-Antoine Jombert, 1749. See also Alexis Claude Clairaut: *Théorie de la figure de la terre – tirée des principes de l'hydrostatique*, Paris: Durand, 1743.

References

Bagrow, Leo. 2017. *History of Cartography*. 2nd ed. London: Routledge.

Bennett, James A. 1986. The Mechanics' Philosophy and the Mechanical Philosophy. *History of Science* 24: 1–28.

———. 1987. *The Divided Circle: A History of the Instruments for Astronomy, Navigation and Surveying*. Oxford: Phaidon Christie's.

———. 1998. Practical Geometry and Operative Knowledge. *Configurations* 6: 195–222.

Brachner, Alto. 1983. *G.F. Brander, 1713–1783: Wissenschaftliche Instrumente aus seiner Werkstadt*. München: Dt. Museum.

Brahe, Tycho. 1972. *Astronomia instauratae mechanica (1598)*. Reprint = Opera omnia, vol. 5. Havniae: Libraria Gyldendaliana.

Buisseret, David. 2003. *The Mapmaker's Quest: Depicting New Worlds in Renaissance Europe*. Oxford: Oxford University Press.

Campbell, J.B. 2000. *The Writings of the Roman Land Surveyors: Introduction, Text, Translation and Commentary*. London: Society for the Promotion of Roman Studies.

Cantor, Moritz. 1875. *Die römischen Agrimensoren und ihre Stellung in der Geschichte der Feldmesskunst*. Leipzig: Teubner.

Chabert, Jean-Luc. 1999. Algorithms for Arithmetic Operations. In *A History of Algorithms*, ed. Jean-Luc Chabert, 7–47. Berlin: Springer.

Chapman, Allan. 1995. *Dividing the Circle. The Development of Critical Angular Measurement in Astronomy 1500–1850*. 2nd ed. Cichester: Wiley.

Collins, Edward. 2014. Portuguese Pilots at the Casa de la Contratación and the Examens de Pilotos. *The International Journal of Maritime History* 26: 179–192.

Coyne, George V., Michael Anthony Hoskin, and Olaf Pedersen, eds. 1983. *Gregorian Reform of the Calendar: Proceedings of the Vatican Conference to Commemorate its 400th Anniversary*. Vatican City: Pontifical Academy of Sciences.

Damerow, Peter, and Wolfgang Lefèvre, eds. 1985. *George Adams: Geometrische und graphische Versuche. Nach der deutschen Ausgabe von 1795*. Darmstadt: Wissenschaftliche Buchgesellschaft.

Dilke, Oswald A.W. 1971. *The Roman Land Surveyors. An Introduction to the Agrimensores*. Newton Abbot: David & Charles.

DMD. 2006ff. *DMD – Database Machine Drawings – by Wolfgang Lefèvre and Marcus Popplow*. Berlin: Max Planck Institute for the History of Science. http://dmd.mpiwg-berlin.mpg.de/home.

Drachmann, A.G. 1969. A Detail of Heron's Dioptra. *Centaurus* 13 (3): 241–247.

Dreier, Adrian. 1979. *Winkelmessinstrumente. Vom 16. bis zum frühen 19. Jahrhundert*. Berlin: Kunstgewerbemuseum Berlin.

Dreyer, John L.E. 1972. *Tycho Brahe: Ein Bild wissenschaftlichen Lebens und Arbeitens im sechzehnten Jahrhundert*. Vaduz: Sänding.

Feldmann, Richard W. 1961. The Cardano-Tartaglia Dispute. *The Mathematics Teacher* 54 (3): 160–163.

Folkerts, Menso. 2014. Die Mathematik der Agrimensoren – Quellen und Nachwirkung. In *In den Gefilden der römischen Feldmesser*, ed. Eberhard Knobloch and Cosima Möller, 131–148. Berlin: De Gruyter.

Goldthwaite, Richard. 1972. Schools and Teachers of Commercial Arithmetic in Renaissance Florence. *Journal of European Economic History* 2: 418–433.

Haasbroek, N.D. 1968. *Gemma Frisius, Tycho Brahe and Snellius and their Triangulations*. Delft: W.D. Meinema.

Haselberger, Lothar. 1999. *Essence and Appearance. Refinements in Classical Architecture*. Philadelphia: University of Pennsylvania Museum Publications.

Høyrup, Jens. 2009. The Rare Traces of Constructional Procedures in "Practical Geometries". In *Creating Shapes in Civil and Naval Architecture: A Cross-Disciplinary Comparison*, ed. Horst Nowacki and Wolfgang Lefèvre, 367–377. Leiden and Boston: Brill.

———. 2019. *Selected Essays on Pre- and Early Modern Mathematical Practice*. Cham: Springer.
Hutton, Charles. 1795. *A Mathematical and Philosophical Dictionary*. London: Johnson and Robinson.
Kern, Ralf. 2010. *Vom Astrolab zum mathematischen Besteck: 15. und 16. Jahrhundert*. Köln: Verlag der Buchhandlung Walther König.
King, David. 2011. *Astrolabes from Medieval Europe*. Farnham: Ashgate Variorum.
Kretschmer, Konrad. 1909. *Die italienischen Portolane des Mittelalters: ein Beitrag zur Geschichte der Kartographie und Nautik*. Berlin: Mittler.
Krücken, Wilhelm. 1994. Die Mercator-Weltkarte AD USUM NAVIGANTIUM 1569. In *Gerhard Mercator – Europa und die Welt. Begleitband der Ausstellung [...] anläßlich des 400. Todestages von Gerhard Mercator*, ed. Ruth Löffler and Gernot Trommau, 211–220. Duisburg: Kultur- und Stadthistorisches Museum Duisburg.
Peters, Gunthild. 2018. *Zwei Gulden vom Fuder: Mathematik der Fassmessung und praktisches Visierwissen im 15. Jahrhundert*. Stuttgart: Franz Steiner.
Price, Derek J. 1957. The Manufacture of Scientific Instruments – from c. 1500 to c. 1700. In *A History of Technology*, ed. Singer Charles, E.J. Holmyard, A.R. Hall, and Trevor I. Williams, 620–647. Oxford: Clarendon Press.
Renn, Jürgen. 2020. *The Evolution of Knowledge – Rethinking Science for the Anthropocene*. Princeton/Oxford: Princeton University Press.
Repsold, Johann Adolf. 1908. *Zur Geschichte der astronomischen Messwerkzeuge I: Von Peurbach bis Reichenbach – 1450 bis 1830*. Leipzig: Engelmann.
Richeson, Allie W. 1966. *English Land Measuring to 1800: Instruments and Practices*. London: Society for the History of Technology.
Saridakis, Voula. 2001. *Converging Elements in the Development of Late Seventeenth-Century Disciplinary Astronomy: Instrumentation, Education, and the Hevelius-Hooke Controversy*. Master of Science, Science and Technology Studies, Virginia Polytechnic Institute and State University.
Schemmel, Matthias. 2008. *The English Galileo: Thomas Harriot's work on motion as an example of preclassical mechanics*, Boston Studies in the Philosophy and History of Science. Vol. 1. Dordrecht: Springer.
Schilder, Günter. 2017. *Early Dutch Maritime Cartography: The North Holland School of Cartography (c. 1580–1620)*. Leiden/Boston: Brill.
Schneider, Ivo. 1970a. *Der Proportionalzirkel. Ein universelles Analogrecheninstrument der Vergangenheit*. München: Oldenbourg.
———. 1970b. Die mathematischen Praktiker im See-, Vermessungs- und Wehrwesen vom 15. bis zum 19. Jahrhundert. *Technikgeschichte* 37 (3): 210–242.
Schmidt, Fritz. 1935. *Geschichte der geodätischen Instrumente und Verfahren im Altertum und Mittelalter*. Neustadt an der Haardt: Pfälzische Gesellschaft zur Förderung der Wissenschaften.
Smith, L. Murphy. 2013, January. Luca Pacioli: The Father of Accounting. *Social Sciences Research Network Electronic Journal*.
Snyder, John P. 1993. *Flattening the Earth: Two Thousand Years of Map Projections*. Chicago: The University of Chicago Press.
Swetz, Frank J. 1989. *Capitalism and Arithmetic*. La Salle: Open Court.
Thoren, Victor E. 1973. New Light on Tycho's Instruments. *Journal for the History of Astronomy* 4: 25–45.
Torge, Wolfgang. 2007. *Geschichte der Geodäsie in Deutschland*. Berlin: de Gruyter.
Van Egmond, Warren. 1981. *Practical Mathematics in the Italian Renaissance: A Catalog of Italian Abbacus Manuscripts and Printed Books to 1600, Supplementi agli Annali dell'Istituto e Museo di Storia della Scienza*. Florence: IMSS.
Verdun, Andreas. 2014. *Leonhard Eulers Arbeiten zur Himmelsmechanik*. Berlin: Springer.
Vogel, Kurt. 1971. Fibonacci, Leonardo. In *Dictionary of Scientific Biography*, ed. Charles Coulston Gillispie, vol. 4, 604–613. New York: Charles Scribner's Sons.

Weißbecker, Bernhard. 2012. *Das Uhrwerk des Mondes: Tobias Mayer und der Längenpreis*. Norderstedt: Books on Demand GmbH.

Wepster, Steven. 2010. *Between Theory and Observations: Tobias Mayer's Exploration of Lunar Motions*. Berlin: Springer.

White, Lynn. 1978. *Medieval Religion and Technology – Collected Essays*. Berkeley/Los Angeles/ London: University of California Press.

Whitfield, Peter. 1996. *The Charting of the Oceans: Ten Centuries of Maritime Maps*. London: British Library.

Zinner, Ernst. 1931. *Die Geschichte der Sternkunde: Von den ersten Anfängen bis zur Gegenwart*. Berlin: Springer.

———. 1967. *Deutsche und niederländische astronomische Instrumente des 11. bis 18. Jahrhunderts*. 2nd ed. München: Beck.

Chapter 8
Epilog

An epilog is expected to provide a resume of a book's findings. In our case, however, such a resume seems almost impossible in light of the broad range and diversity of facts and complex relationships described and discussed in the six chapters. But we can recapitulate the different patterns of interrelations between learned and practical knowledge we found to be characteristic of the four types of technologically advanced fields of praxis distinguished in the introduction – mechanics, practical mathematics (practical astronomy, mathematical geography and cartography, navigation, and surveying), chemistry, and "conglomerate" sciences (sciences of architecture and mining). Based on this recapitulation, we can eventually draw some conclusions on the implications of our findings for the understanding of the early modern history of knowledge, including the Scientific Revolution.

8.1 Mechanics

The characteristic pattern of interrelations between learned and practical knowledge exhibited by the field of *Mechanics* – mechanics and mechanical engineering – can be summarized by the following four steps: (1) Almost no relations up to the mid-sixteenth century; (2) ambivalent contacts around 1600; (3) largely separate paths of development during the seventeenth century; (4) in the course of these largely independent developments, both sides generated prerequisite conditions for fruitful interactions that commenced in the eighteenth century.

Some key words or cues may recall our findings regarding these four steps.

(1) The medieval science of mechanics largely ignored the practical mechanics of the age as well as the illuminated manuscripts on machines that emerged around 1400. In turn, medieval practical mechanics did not take notice of the Latin manuscripts on mechanical issues circulating in the distinguished sphere of the universities. The only exception was

W. Lefèvre, *Minerva meets Vulcan: Scientific and Technological Literature – 1450-1750*, Archimedes 60, https://doi.org/10.1007/978-3-030-73085-7_8

Vitruvius' short chapters about machinery which some engineers already noticed in the fifteenth century.

(2) The revival of studies in mechanics in the second half of the sixteenth century was due not to questions and problems posed by contemporary technological literature on mechanical engineering but to the rediscovery of classical texts (Archimedes, Hero, and the Ps-Aristotelian *Quaestiones mechanicae*). These studies found some resonance among mechanical engineers but proved to be of little help, particularly because they initially excluded dynamical issues. In the case of gunnery, we encountered another kind of contact between learned and practical mechanics, in particular a classical example of a practical problem – the range and trajectory of shots – that became a challenging object for learned mechanics. However, the ensuing theoretical ballistics largely developed detached from practical ballistics and assumed the character of "shooting with ink," as Jochen Büttner put it. Thus, both kinds of contacts between scientific and practical mechanics were remarkably ambivalent in so far as they widened rather than narrowed the gap between practical and the incipient modern scientific mechanics, which was a science of idealized (geometrical) subjects rather than material ones.

(3) In the seventeenth century a certain stagnation could be observed as regards the technological literature on mechanical engineering. Yet technical developments continued – key word *Machine de Marly* – not because of a better theoretical understanding of machines but because of engineers' rich practical experience. At the same time, engineers became increasingly aware of the limitations of engineering on this basis when faced with the urgent need to calculate the performance of advanced and extended machinery. The science of mechanics, which developed nothing less than the basic concepts of modern (Newtonian) dynamics in the seventeenth century, owed almost nothing to the contemporary technological literature on mechanical engineering. There is, however, one important exception: practical mechanics, in fact, constituted an essential basis for the novel scientific method of studying mechanical questions by experimenting. Moreover, in the case of gunnery, it was experimentation, devised and performed by practitioners, that gave rise to what is known as the Ballistic Revolution of the eighteenth century.

(4) Eventually, in the eighteenth century, the new science of mechanics, along with the new tool of mathematical analysis, made it possible to tackle some of the urgent problems of mechanical engineering – measurement of the driving forces of machines, calculation of friction, finding optimal shapes for machine parts like cogs, and so on. This advanced science not only met up with competent mechanical engineers but also with an advanced technology that allowed to appropriate and translate useful scientific results. As a result, a new and close relationship between technological and scientific literature on mechanics arose and continued developing.

Probably the most striking feature of this pattern of interrelations between learned and practical knowledge is the separation of the developmental paths of these two kinds of knowledge during most of the early modern period. It persisted up to the eighteenth century despite the repeated contacts and mutual awareness of learned men and practitioners, and even continued despite challenging objects like, for example, the trajectory of projectiles which became even the subject of main parts of Galileo's *Discorsi* as well as Newton's *Principia*. True, these objects provoked revisions of natural philosophical assumptions; but they failed to stimulate investigations into the realities of gunnery. Thus, the seventeenth-century elaborations of basic conceptions of modern (Newtonian) mechanics cannot be taken as resulting from reflections on the procedures, tools, or the insights and open questions of contemporary engineers or gunners. Rather, these elaborations were bound to follow a particular style of reasoning they inherited from ancient models: the style of deductive (Euclidean) reasoning by mathematical means. The pattern of interrelations and interactions between learned and practical knowledge in early modern mechanics owes much to exactly this style of reasoning.

Deductive reasoning implied not only certain methodological constraints but also imposed an idealization of the subjects of inquiry: physical objects such as levers or projectiles then became geometrical lines or points. Idealization proved to be a serious obstacle for interactions between scientific and practical mechanics. It was a major reason why practical mechanics was unable to benefit from scientific mechanics for a long period. What is more, this science of mechanics failed when confronted with urgent practical problems such as friction, recoil or aerodynamic drag. Idealization was also a major reason why real levers or projectiles were irrelevant for conceptualizations in this style of deductive reasoning. On the other hand, there was a latent and at times even open conflict between the deductive style of reasoning and the orientation of the incipient modern mechanics, which was empirical in principle. The solution of this conflict required special access to the experiential world, namely access mediated by experiments, particularly those that procured exact measurements as an operational basis for mathematical reasoning. It was exactly this constellation in which the achievements of practical mechanics became a resource of scientific mechanics, namely in relation to its experimental work.

8.2 Practical Mathematics

The characteristic pattern of interrelations between learned and practical knowledge exhibited by the field of *Practical Mathematics* can be summarized by the following features: (1) All of the various domains of practical mathematics were related to one or several mathematical theories; (2) in some of these domains, only the results of certain mathematical theories were used pragmatically, that is, without considering the derivation or justification of these results by the pertaining theory; (3) other domains dealt with practical problems that presupposed the framework of a mathematical theory and could only be solved in this framework; (4) as regards the question of any possible feedback the mathematical theories might receive from being used or applied in the domains of practical mathematics, it makes a considerable difference whether such a domain operated in the framework of a mathematical theory or only used its results pragmatically.

Some key words or cues may again recapitulate our findings regarding these four features:

(1) Early modern surveying, as well as higher geodesy, naturally related to geometry. Mathematics for practitioners (merchants, craftsmen, artists, and so on) as taught by practical mathematicians (computing or reckon masters) comprised fragments of learned arithmetic and elementary geometry. Practical astronomy and mathematical geography presupposed and operated in the framework of the mathematical model of heaven and earth the West had inherited from Greek and Hellenistic Antiquity through the writings of Ptolemy.

(2) Practical surveying in the realm of civil or military engineering as well as of mining explicitly or (more usually) implicitly used Euclidean theorems, such as that of intersecting

lines, regardless of their justification or function in Euclid's theory building. This pragmatic and eclectic use of suitable results of learned mathematics was characteristic of all fields of practical geometry.[1]

(3) The standard tasks of early modern practical astronomy – determining the beginning and end of the seasons and particularly the Eastern date, calculating ephemerides, and creating horoscopes – were intelligible only in the framework and coordinate system of a mathematical model of celestial movements like that of Ptolemy, Brahe, or Kepler, and could only be performed by using the relevant mathematics. The same holds for the early modern mathematical geography and cartography which built upon and further developed the mathematical model of the Earth inherited from Ptolemy.

(4) Learned mathematics remained largely untouched by developments in practical surveying or other domains of practical geometry, at least up to the seventeenth century when new mathematical disciplines arose, inspired by certain constructive geometrical techniques used by architects and painters. In contrast, developments in practical astronomy and geography led to fundamental revisions of the underlying mathematical model of celestial movements and thereby eventually to the destruction of the classic cosmology inherited from Greek Antiquity.

These features of the interrelations between technological and scientific knowledge in the various fields of early modern practical mathematics are artificial to some extent, because in some fields of practical mathematics technological and scientific literature, and thus technology and science, appear indistinguishable. Particularly in practical astronomy, geography and, later, higher geodesy, literature can be found that is technological and scientific alike. This is due to the dual nature of many activities of practical mathematicians: they pursue a practical goal and contribute, at the same time, to the development of science. Improved astronomical observations in the context of calendrical calculations, for example, enrich the empirical knowledge about celestial movements and therefore involve the tacit empiric presuppositions of the mathematical model of these movements; in the long run, this potentially could and actually did lead to revisions of this model. Thus, explorations and calculations undertaken for practical goals in the framework of such a mathematical model are not just applications but substantiations or challenging confrontations of this model by and with the empirical world. Accordingly, the literature on techniques and performances of practical mathematicians is often not just technological literature in relation to scientific literature but an empirical complement to the latter.

---

[1] This pragmatic and eclectic commerce between learned and practical geometry was not one-sided, as exemplified by the interrelations between two sides of Euclid's geometry itself. In Euclid's framework, constructions, e.g. dropping a perpendicular onto a straight line or bisecting a given angle, play an essential and indispensable role, though a serving, auxiliary one. They procure evidences on which the theorems can base their deductions. Out of the wealth of geometrical constructions employed in Greek practical geometry – think, for example, of the refinements used in architecture – Euclid selected only a few constructions that were of use for the deductive purposes of his work. Subsequently, practitioners, for their part, selected and used theorems of Euclid's geometry that suited their purposes, such as the proportionality of the lengths of sides of similar triangles. They did so without caring about how these results had been proved or deduced by Euclid or subsequent theoretical geometricians. Here, it was the theorems' turn to play an auxiliary role. See Lefèvre (2019).

8.3 Chemistry

The characteristic pattern of interrelations between learned and practical knowledge exhibited by the field of *Chemistry* can be summarized by the following features: (1) The learned counterpart of early modern chemical practices was not a unified body of chemical theories but a few diverse natural philosophical theories of the ultimate constitution of matter. (2) As regards specific chemical issues, no learned or theoretical literature existed separately from the technological literature before the seventeenth century. Rather, theoretical explanations and practical descriptions and instructions were, as a rule, arranged in one and the same booklet or manual. (3) The relationship between natural philosophical assumptions and chemical practices was one of associations and analogies. The two sides did not build upon one another. (4) A new theory of chemical processes arose in the decades around 1700. This theory was not a further development of existing natural philosophical theories of matter but the result of reflections on the behavior of chemical substances in certain chemical processes as described in the associated scientific-technological chemical literature.

Some key words or cues may again recall our findings regarding these four features.

(1) The West had inherited, among a host of barely elaborated so-called alchemistical "theories" of various origins, a number of ancient Greek philosophical theories of the ultimate constitution of matter – an Aristotelian, a neo-Platonic, and a Stoic theory besides several atomistic theories. The reception of this heritage by early modern chemists cannot be called anything but eclectic. This holds particularly for Paracelsus and his followers.

(2) Before the eighteenth century, technological and scientific writings are mostly indistinguishable in the field of chemistry. Casual statements about or deliberate and explicit expositions of theoretical assumptions held by an author were part and parcel of almost every significant technological work on chemistry since Brunschwig's *Destillierbuch* of 1500. In the seventeenth century, it had become the standard for chemical manuals to begin with a theoretical part before going on to the various technical descriptions and instructions.

(3) However, as we found, these two parts were largely autonomous, even though the theoretical and practical explanations could be linked with each other: Theoretical terms like extraction, sublimation or distillation, for example, were referring to technical processes and at the same time utilized to lend some empirical plausibility to assumptions in the framework of natural philosophical theories of matter, that is, to theories that had originated and developed completely independently of chemical technologies. This holds not only for the inherited ancient philosophical theories but also for early modern theories of the ultimate constitution of matter, the Paracelsian ones as well as the atomistic matter theories of the seventeenth century. And while the categories and key assumptions of these philosophical theories of matter owed nothing to developments of chemical procedures or the expansion of the multitude of substances they processed, conversely, these technological developments owed nothing at all to these theories.

(4) Around 1700 a new type of theory began to emerge. It was not another philosophical matter theory yet again, but a "chemical theory" in the modern sense of the term, that is, a theory of the laws that rule the behavior and interactions of chemical substances. This theory resulted from reflections of metallurgical processes and above all of new procedures of salt production – reflections based on a rich technological literature on these processes and procedures.

Apparently two different relationships occurred between learned and practical knowledge in the field of early modern chemistry: first, up to the seventeenth century, the relationship between practical knowledge and natural philosophical theories of matter and second, from the mid-seventeenth century on, the relationship between practical knowledge and an emerging theory of certain interactions of chemical substances.

The first relationship, that between practical knowledge and natural philosophical theories, proved to be bewilderingly paradoxical, and this must be stressed once more: On the one hand, both sides used the other as a kind of resource – the technological literature used philosophical doctrines when interpreting technical processes such as distillations and, conversely, philosophical explanations used such processes as illustrations and examples of their statements. On the other hand, the two sides did not impact on one another – technological developments did not owe anything to the philosophical theories of matter, and nor did technological developments have any effect on the doctrines of these philosophical theories. This also holds for the attempts of some seventeenth-century chemists to conceptualize chemical processes based on resurrected atomistic or newly developed corpuscular theories: These attempts were not a reaction to new technological developments – and, by the way, led nowhere.

What eventually prompted a new theory of chemical processes was indeed certain technical developments, or more precisely, irritating occurrences brought about by these developments or challenging objects such as the reversibility of certain chemical processes, the preservation of substances in dissolutions, and so on. However, it was not interpretations of these technical developments in terms of ultimate causes as proposed by one of the philosophical matter theories that led to this new chemical theory but reflections focused on intermediate and experimentally accessible causes.[2] In its beginnings, this new theory profited less from experiments than from systematic analyses and evaluations of relevant technologies as described in the rich technological chemical literature of the seventeenth century. The close relationship between technological and scientific literature is rarely as clear as in this case, where a scientific theory evolved through studying and conceptualizing natural laws as utilized in technological processes and documented in technical manuals.

[2] For the notions "ultimate causes" and "intermediate and experimentally accessible causes," see Chalmers (2012). Chalmers had discussed this topic earlier with specific regard to Robert Boyle's attempt at conceptualizing chemical processes on the basis of a corpuscular theory; see Chalmers (2010).

8.4 Architecture and Mining

The characteristic pattern of interrelations between learned and practical knowledge exhibited by the sciences of *Architecture* and *Mining* can be summarized by the following features: (1) The principal structure of the various fields of knowledge of each of these two "conglomerate sciences" was already delineated by scholarly literature at the start of their career in the early modern period. (2) In both cases, not all of the various fields were interconnected with counterparts in the world of learning. Some fields had relations with mathematics and mechanics that were similar in principle to those observed in the fields of practical mathematics and mechanical engineering. But relationships with other fields of learning also existed. (3) In both cases, some fields of practical knowledge became points of departure for new scientific disciplines. (4) In both cases, the relations between learned and practical knowledge were institutionalized by the foundation of special academies in the eighteenth century.

Some key words or cues may again recall our findings regarding these four features.

(1) The delineation of the "disciplines" of the science of architecture as given by Vitruvius' *De architectura libri decem* became canonical for the early modern literature on architecture following Alberti's *De re aedificatoria* (ca. 1450). Agricola's *De re metallica* (1556) had the same canonical function as regards the fields of knowledge of mining science in the early modern period.

(2) Both sciences comprised fields of manual know-how such as that of masons or carpenters, practical knowledge without relationships to learned knowledge. Both sciences comprised fields of practical knowledge with relationships to the classical fields of learned knowledge, that is to scientific mechanics (e.g., in the case of architecture, structural design or, in the case of mining, mechanical engineering of drainage systems, etc.) and to mathematics (surveying, drafting, etc.). Both sciences had also relationships to more remote fields of learned knowledge, e.g., in the case of architecture, to ancient, archeological, and aesthetic learning or, in the case of mining, to metallurgy and mineralogy. Furthermore, both sciences comprised fields of expert knowledge with no counterpart in the world of learned knowledge, e.g., in the case of architecture, knowledge about construction materials and quarries or, in the case of mining, knowledge about deposits of ores and other minerals.

(3) In both cases, developments of certain kinds of practical knowledge and techniques led to the emergence of new scientific disciplines. In the case of architecture, techniques of practical (constructive) geometry, as embodied in linear perspective and stereotomy, became starting points for new non-Euclidean mathematical disciplines; in mining, systematic reflections of the accumulated practical knowledge about ore deposits, mountain formations, and stratigraphic facts opened up the era of modern geology.

(4) Special educational institutions for structural engineering and mining which combined practical and theoretical instruction were founded in eighteenth-century France, Prussia, and the Hapsburg monarchy. These institutions correspondingly inspired publication of textbooks and other teaching materials which offered scientific as well as technological teaching content.

Perhaps the most striking feature of the interrelations between practical and learned knowledge in the case of the sciences of architecture and mining is that one encounters the whole range of possible relationships – from none at all up to those of mutual support. This is due to the heterogeneity of the various kinds of knowledge that are utilized by these "conglomerate sciences." And this implies that these sciences remained structured by different kinds of knowledge stemming from several cooperating professions rather than by an overarching theory, even if some of their fields of knowledge developed a very close relationship to scientific knowledge. But this probably holds true for all of the technological sciences and not just for architecture and mining alone.

8.5 Interrelations and Developments

This recapitulation of our investigations of the interrelations between the early modern technological and scientific literature in six different fields of knowledge and practice displays not only the wide range but also the main features of these interrelations:

(1) in the case of mechanics and ballistics, the main feature is that challenging technological objects first triggered internal developments of a purely theoretical nature on the scientific side before real exchanges between the two sides became possible; (2) in the case of practical astronomy and practical mathematical geography, the main feature is that technologies operating in the framework of theoretical models substantiated but also challenged these models' reference to reality; (3) in the case of chemistry, the salient feature is that for a long period, technological developments only had associative relations with natural philosophical theories before some of them triggered the emergence of a new scientific theory.

At the same time, these main features or types of interrelations indicate major ways in which sciences and technologies developed in the early modern period – *development paths* which should be briefly outlined in concluding this epilog.

As regards the *development of technologies*, one can generally state that their admirable achievements owed next to nothing to traditional or newly developing scholarly theories before the eighteenth century. One exception is the role played by the inherited theories of statics and mathematics, particularly geometry, in a considerable number of practical fields. Results of these theories were used in an eclectic and pragmatic way by practitioners in these fields, as we have seen. Another exception is the development of mathematical instruments, which owed much to mathematics (for the gradations (divisions) of the instruments) as well as to optics. The development of these instruments was decisive for improvements in practical astronomy, mathematical geography, navigation and cartography, as well as lower and higher geodesy, that is, for practical fields on the borderline to theoretical disciplines. It is therefore not surprising that these developments frequently led to the invention of instruments for specific scientific purposes – observation, measuring, and experimentation.

The development of the incipient *science of mechanics* is often taken as paradigmatic for the genesis of modern sciences. However, it is frequently overlooked that its precarious relations to practical mechanics manifest a profound ambivalence regarding the theoretical constitution of this science in the early stage of its development. It started as a deductive (Euclidean style) enterprise with vague or at least barely substantiated claims to empirical validity. Occupied with theoretical revisions and reconstructions of traditional conceptions of dynamics, it initially gave rise to theorems such as the parabolic trajectory of projectiles that testified to achievements in dynamics but could not stand the test of empirical findings or practical applicability. No wonder practical mechanics could not utilize such theorems or enlist the support of this science of mechanics in overcoming its bottlenecks. And conversely, the early modern science of mechanics found little more in contemporary practical mechanics than thought-provoking problems, with no starting points for promising conceptualizations.

Thus, before the eighteenth century, theoretical and practical mechanics developed largely along separate paths. However, each of them on its own path generated achievements that eventually proved to be prerequisite conditions for productive relations in which each could profit from the developments of the other. On the theoretical mechanics side, it was particularly new powerful mathematical methods (above all mathematical analysis) that made it possible to tackle intricate empirical problems like friction, aerodynamical drag, or recoil; and on the practical mechanics side, it was standardized mechanical devices and precision instruments that facilitated the use of theoretical mechanics for practical purposes and opened up access to the empirical world for this science by means of experiments. The latter enabled the science of mechanics to become a real empirical theory.

With regard to practical mathematics, the development of *astronomy* – and, connected with it, of mathematical geography – deserves particular attention. The development of early modern astronomical theory – from Copernicus and Kepler to Newton – is well known, along with the fact that practical mathematicians contributed to this development with observations, observational instruments, and so on. The interaction of theoretical and practical astronomy in this development was, however, not entirely simple and is therefore worth briefly recapitulating.

The astronomy inherited from classical Antiquity and the Middle Ages was beset with two major contradictions – the lack of congruency between the philosophical cosmology and the Ptolemaic mathematical model of the celestial movements on the one hand, and, on the other, the imperfect congruency between this model and the observed celestial movements. Copernicus tackled the first of these contradictions: He proposed a mathematical model that conformed to the main cosmological doctrine that celestial bodies make circular orbits with uniform motion around the cosmological center (albeit the sun instead of the Earth) and not around centers of epicycles, as in Ptolemy's model. This proposal was internal to theory and had no connection with practical astronomy. The second of these contradictions increasingly concerned practical astronomers of the fifteenth and sixteenth centuries and eventually led to the reformed Gregorian calendar. This spurred further efforts to obtain more accurate observations by means of improved observational instruments – endeavors by figures like Tycho Brahe who were both practical and theoretical astronomers. The resulting revision of the Copernican model by Kepler – who posited elliptical instead of circular planetary movements with varying velocity – initiated the process which led to abandoning traditional cosmology in favor of modern mathematical and physical astronomy. In this "revolution," practical mathematics, in this case practical astronomy, definitely had a share; it represented not only the practical dimension of the new astronomy but also an essential part of its empirical dimension.[3]

It is important to emphasize the empirical character of the new astronomy because it marks a significant difference between this incipient modern science and that of mechanics. Modern astronomy did not begin as a deductive science by revising and reconstructing an inherited conceptual framework. Rather, Kepler's three laws of planetary motion, which allowed the generation of Hooke and Newton to transform astronomy into a physical science, were not derived from theoretical principles – they were found through a painstaking analysis of observational data. In this respect the beginning and early stage of modern astronomy resembled the beginning and early stage of modern chemistry more than that of mechanics.

A *chemical theory* in today's meaning of the term, that is, a theory based on knowledge of laws that rule the interactions of chemical substances, emerged in the decades around 1700 through comparative analyses and reflections of certain chemical operations in metallurgy and the brand-new production of salt by dissolution and precipitation. This science, too, started as an empirical one – by induction rather than deduction, to put it in terms of philosophy of science. And comparably to the abandonment of ancient philosophical cosmology by the new astronomical theory, this new chemical theory eventually led to the abandonment of theories from natural philosophy of the ultimate constitution of matter as a frame of reference for reflections. In both cases, explanations shifted the focus from an intangible ontological sphere to accessible causes – a "key to understanding the scientific revolution" as Alan Chalmers put it.[4]

There is, however, an important difference between the empirical investigations at the beginning of the new sciences of astronomy and chemistry. The observational data studied by the former were already geometrically encoded within the celestial coordinate system and could therefore be analyzed by mathematical means. This was not an option in chemistry, where reaction patterns of chemical substances had to be compared, weighed, and ordered to discover the laws governing these patterns. In this respect chemistry resembled the incipient "Baconian sciences" of magnetism or of electricity rather than astronomy.[5]

Considering the variety of ways in which the modern physical sciences developed in the early modern period, it appears highly questionable for an historian to privilege one of these ways as epitomizing the Scientific Revolution. Only an integrated view or synopsis of this variety can do justice to both the richness and the contingencies of this development. The same holds for the development of

[3] This role of practical mathematics in the development of a scientific theory can also be observed in the closely related case of mathematical geography and cartography, which outgrew the Ptolemaic model on the basis of the empirical measurement of the Earth by navigators and geodesists.

[4] Chalmers made this point both with regard to futile atomistic attempts to explain chemical processes in the seventeenth century and to hydrostatical theories of that century; see Chalmers (2017).

[5] For the notion of ‚Baconian sciences', see Kuhn (1976) 10ff.

technologies in this period and particularly for the technological literature which has taken center stage in this book. There was no royal road to the modern physical sciences or to modern technology.

References

Chalmers, Alan F. 2010. Boyle and the Origins of Modern Chemistry: Newman Tried in the Fire. *Studies in History and Philosophy of Science* 41: 1–10.

———. 2012. Intermediate Causes and Explanations: The Key to Understanding the Scientific Revolution. *Studies in History and Philosophy of Science* 43: 551–562.

———. 2017. *One Hundred Years of Pressure: Hydrostatics from Stevin to Newton*. Cham: Springer.

Kuhn, Thomas S. 1976. Mathematical vs. Experimental Traditions in the Development of the Physical Science. *The Journal of Interdisciplinary History* 7 (1): 1–31.

Lefèvre, Wolfgang. 2019. Drawing Instruments. In *Culture and Cognition: Essays in Honor of Peter Damerow*, ed. Jürgen Renn and Matthias Schemmel, 161–165. Berlin: Edition Open Access Max Planck Institute for the History of Science.

technologies in this period and particularly for the technological literature which has taken center stage in this book. There was no revolution [illegible]

References

[illegible]

Picture Credits

Bayerische Staatsbibliothek, München (BSB)
3.1, 4.1, 5.1
Library of the Max Planck Institute for the History of Science, Berlin
2.2, 2.3, 2.4, 2.5, 3.2, 3.3, 4.2, 4.3, 4.4, 4.5, 5.2, 5.3, 5.4, 5.5, and 5.7, 6.1, 6.2, 7.2
The Metropolitan Museum of Art, New York
2.1
Wikipedia – Public Domain
5.6, 6.1, 7.1

W. Lefèvre, *Minerva meets Vulcan: Scientific and Technological Literature – 1450-1750*, Archimedes 60, https://doi.org/10.1007/978-3-030-73085-7

Index

W. Lefèvre, *Minerva meets Vulcan: Scientific and Technological Literature – 1450-1750*, Archimedes 60, https://doi.org/10.1007/978-3-030-73085-7

The manufacturer's authorised representative in the EU is Springer
Nature Customer Service Centre GmbH, Europaplatz 3, 69115 Heidelberg,
Germany. If you have any concerns regarding our products, please
contact ProductSafety@springernature.com

Printed and bound by CPI Group (UK) Ltd, Croydon, CR0 4YY
30/09/2026
02225505-0004